MOLECULAR ACOUSTICS

MOLECULAR ACOUSTICS

ANDREW J. MATHESON

Lecturer in Chemistry
University of Essex

WILEY–INTERSCIENCE
a division of John Wiley & Sons Ltd
London · New York · Sydney · Toronto

Library of Congress catalog card number 73–147401

ISBN 0 471 57620 4

Printed by John Wright & Sons Ltd, at the Stonebridge Press, Bristol, England

PREFACE

This book describes the molecular information which may be obtained from studies of the propagation of high-frequency longitudinal and shear waves in gases, liquids, polymers and solids. It is written at a level suitable for the graduate starting research work, although some of the material has been used in teaching final-year undergraduates in Chemical Physics at the University of Essex.

I am indebted to the following organizations for permission to reproduce illustrations:

Academic Press: Figure 8.1.1.

Acoustical Society of Japan: Figures 14.2.2 and 14.3.1.

Acustica: Figures 4.5.1, 4.5.2 and 14.2.1.

American Ceramic Society: Figure 9.4.3.

American Chemical Society: Figures 7.4.2, 7.4.6, 7.4.7 and 13.3.1.

American Institute of Physics: Figures 3.2.1, 3.8.1, 3.8.2, 4.5.3, 4.6.1, 6.3.1, 6.5.1, 6.10.1, 7.4.1, 10.4.1, 10.5.1, 10.6.2, 10.7.1, 10.7.2, 11.3.1, 11.4.1, 12.4.8, 12.6.2 and 13.1.1.

Berichte der Bunsen-Gesellschaft für Physikalische Chemie: Figures 12.4.4, 12.4.5, 12.4.6 and 12.4.7.

Dr. Dietrich Steinkopff Verlag: Figure 13.2.1.

Faraday Society: Figures 4.8.2 and 14.4.1.

Institute of Physics and The Physical Society: Figures 3.3.1, 6.8.3 and 11.2.1.

Japanese Journal of Applied Physics: Figure 3.4.1.

John Wiley: Figure 7.2.2.

Laboratory Practice: Figures 6.6.2 and 6.7.1.

North Holland Publishing Company: Figures 4.5.4, 8.5.1 and 8.5.2.

The Royal Society: Figures 4.8.1, 6.9.1, 8.3.1, 8.3.2, 8.3.3, 9.4.1, 9.4.2, 11.2.2 and 12.5.2.

I am grateful to Miss P. J. Tweed and Mrs. B. Zukowski for typing the manuscript of this book.

A. J. MATHESON
Colchester
6th August, 1970

CONTENTS

LIST OF SYMBOLS

Although it has sometimes been necessary to use the same symbol with two or more different meanings, the context should make the identification clear.

a	$C_{v\infty}/C_v$ (2.5.7)
a	radius of crystal (3.1, 6.6)
a	amplitude of vibration (3.4)
a	sum of molecular radii (4.3)
a	geometrical constant (6.4)
a	$\phi/N\zeta$ (7.2)
a_0	initial amplitude of vibration (3.4)
a_0	amplitude of ultrasonic wave (14.6)
a_T	shift factor (7.3.5)
A, A', A''	constants
A	area (6.5)
b	$C_{p\infty}/C_p$ (2.5.8)
b	$I/\mu a^2$ (4.3)
b	width parameter in Gaussian function (5.5.8)
b	length of repeat unit (7.2.16)
B, B', B''	constants
B	magnetic flux density (6.4)
B	segmental mobility (7.2)
c	mean molecular translational velocity (11.4)
c	concentration (14.2)
c_{11}, c_{12}, c_{44}	elastic constants (15.1)
C	Sutherland constant (11.4)
C_1	contribution of ν_M to C_v
C_i	$C_v - C_{v\infty}$
C_i^ω	effective value of C_i at frequency ω
C_p	molar specific heat at constant pressure
$C_{p\infty}$	effective value of C_p at 'infinite' frequency
C_p^ω	effective value of C_p at frequency ω
C_v	molar specific heat at constant volume

$C_{v\infty}$	effective value of C_v at 'infinite' frequency
C_v^ω	effective value of C_v at frequency ω
C_0	static capacitance of torsional crystal (6.6)
d	diameter of vessel (3.4)
d	grain diameter (15.3)
D	rate of decay (3.4)
e_B	back e.m.f. (6.4)
E	energy density (3.3)
E	activation energy for viscous flow (9.1)
E	internal energy (9.3)
E_h	energy to create a hole (9.3)
f	frequency
f_c	characteristic frequency, $1/2\pi\tau$
f_r	reference frequency (7.3)
f_R	resonance frequency (6.6)
$f(v_r)$	distribution function of rotational velocities (4.4)
f'_{10}, f'_{01}	probability per molecule per second of collisional de-excitation or excitation of the lowest vibrational mode of a molecule
$f(\tau_s)$	distribution of retardation times (5.3)
F	a function
F	force
$F(\tau_B)$	distribution of structural relaxation times (10.2)
F_ω	value of function F at frequency ω
g	degeneracy (12.3.62)
$g(r)$	radial distribution function
$g(\tau_s)$	distribution of relaxation times
G	free energy

G^*	complex rigidity modulus	l	length of chain segment (7.1)
G'	real part of G^*		
G''	imaginary part of G^*	l_f	mean free path (11.4)
G_0	value of G' at low frequencies	L	signal amplitude (3.4)
G_∞	value of G' at 'infinite' frequency	L	length of rigid molecule (7.2)
G_j	contribution of process j to G_∞ (5.5.1)	L_C, L_N	length of dislocation loop (15.4)
$(G_\infty)_0$	G_∞ at T_0	m	reduced mass of colliding species (4.2)
h	Planck constant	M	molecular weight
$\hbar$	$h/2\pi$ (4.2)	M	reduced mass of oscillator (4.2)
h	hydrodynamic interaction parameter (7.2)	M	mass of moving system (6.4)
H	enthalpy	M^*	complex longitudinal modulus
i	$(-1)^{\frac{1}{2}}$		
i	current	M'	real part of M^*
I	moment of inertia	M''	imaginary part of M^*
I	sound intensity (2.2)	M_0	value of M' at low frequencies
I	integral (4.2.4)		
J	rotational state	M_∞	value of M' at 'infinite' frequency
J^*	complex compliance	M_S	adiabatic value of M^*
J'	real part of J^*	M_T	isothermal value of M^*
J''	imaginary part of J^*	$\bar{M}_n$	number average molecular weight
J_0	value of J' at low frequencies	$\bar{M}_w$	weight average molecular weight
J_∞	value of J' at 'infinite' frequency	$(M_j)_W$	mass of bound water per mole of j ions (14.6)
$(J_\infty)_0$	J_∞ at T_0	$(M_j)_h$	molecular weight of solvated j ions (14.6)
$J(t)$	creep compliance		
$J_r(t)$	recoverable compliance	n	number of particles or molecules
k	rate constant		
k	Boltzmann constant	n	order of diffracted image (3.5.1)
k_r	torque per unit deflexion (6.5)	n	ratio of inner to outer tube radii (6.7)
k_s	calibration constant (3.3.3)	n	number of polymer molecules per unit volume (7.2)
k_t	tube constant (3.2.1)		
K	equilibrium constant	n_1	number of vibrationally excited molecules in unit volume (3.9)
K	width parameter (9.4.5)		
K_1, K_2	constants of torsional crystal (6.6)		
K^*	complex bulk modulus	n^0	initial number of molecules
K'	real part of K^*	n^e	equilibrium number of molecules
K''	imaginary part of K^*		
K_0	value of K' at low frequencies	n^ω	actual number of molecules at frequency ω
K_∞	value of K' at 'infinite' frequencies		
K_2	$K_\infty - K_0$		
l	length		

n_H	number of hydroxyl groups per molecule (9.3.14)	s	distance of travel (6.7)
N	Avogadro number	S	entropy
N	total number of molecules	S_M	ratio of force to displacement (6.4.6)
N	number of polymer chain segments	t	time
N_H	number of hydrogen bonds per mole (9.3)	t_+, t_-	transport numbers (14.6)
$(N_H)_0$	N_H at T_0 (9.3)	T	temperature in Kelvins
$\bar{N}_0, \bar{N}_1$	average number of molecules in ground/first excited vibrational level	T_c	critical temperature
		T_g	glass transition temperature
		T_0	value of T_g in an experiment of infinite duration
$N(\omega_s)$	distribution of relaxation frequencies (5.6.9)	T_r	reference temperature
p	mode number	U	longitudinal phase velocity
p_{10}	probability of de-excitation of first vibrational level in a collision	U^*	complex longitudinal velocity
p_{nm}	probability of transition from state n to state m in a collision	U_c	longitudinal velocity at the critical point
p_{rad}	probability of a radiative transition (3.9)	U_{LJ}	Lennard–Jones potential energy (9.3)
P	pressure	U_M	molar longitudinal velocity (10.5)
P	probability (7.1)	U_0	value of U at low frequencies
P_0	initial pressure		
q_i, q_n	defined in (4.2.19)	U_s	phase velocity of shear waves
Q	excess absorption divided by the number of molecules	U_s^*	complex shear velocity
		U_t	velocity of torsional waves (6.7)
r	distance	U_{tube}	velocity of longitudinal waves in a tube (3.2)
r	radius (3.2)	U_∞	value of U at 'infinite' frequency
r	ratio of capacitances (6.6.1)		
r_0	distance of closest approach	v	velocity
R	gas constant	v	relative velocity (4.2)
R	resistance	v	specific volume
R	modulus of R^* (6.8)	v^*	value of velocity at maximum of p_{10} (4.2)
R^*	complex reflexion coefficient (6.8)	v^*	volume of a molecule (9.3.4)
R'	$C_p - C_v$	v_f	specific free volume
R_c	real part of Z_c (6.7)	v_r	peripheral velocity of rotation (4.4)
R_i	resistance arising from internal friction	v_0	specific volume at T_0
R_L	real part of Z_L	V	molar volume
R_m	real part of moment of impedance (6.5)	V	intermolecular potential, (4.2, 12.4)
R_s	shear mechanical resistance of a solid	V_f	molar free volume
		$\bar{V}_j$	ionic partial molal volume of jth ion (14.6)
R_s	anisotropy factor (15.3)	$(V_j)_h$	molar volume of solvated j ions

V_{mn} matrix element of inter-molecular potential (4.2)

V_S molar volume of solid

V_0 molar specific volume at T_0

W_+, W_- apparent molar mass of ions

x Cartesian coordinate

x distance

x mole fraction (4.8)

x_{10} displacement of oscillator (4.2)

X variable (12.3.9)

X_c imaginary part of Z_c (6.7)

X_L imaginary part of Z_L

X_m imaginary part of moment of impedance (6.5)

X_S shear mechanical reactance of solid

y Cartesian coordinate

y extent of reaction (12.3)

y_0 instantaneous equilibrium extent of reaction (12.3)

Y variable (12.3.9)

z Cartesian coordinate

z distance along tube (3.3)

z $\log(\tau_s/\tau_s')$ (5.5.8)

z ionic charge (14.6)

z_n number of nearest neighbours (9.3.5)

z $2 \cdot 6\pi f n_s / P$ (2.2.9)

Z collision number

Z impedance

Z impedance of coil in motion (6.4)

Z degree of polymerization (8.1)

Z variable (12.3.9)

Z_c shear mechanical impedance of liquid at cylindrical surface (6.7)

Z_L shear mechanical impedance of liquid

Z_M mechanical impedance (6.4)

Z_0 impedance of coil at rest (6.4)

Z_Q shear mechanical impedance of fused quartz (6.8)

Z_r rotational collision number

Z_t total impedance of moving system (6.5)

Z_{10}' number of collisions for energy transfer to ν_M

α absorption coefficient

α parameter in potential energy function (4.2.8)

α logarithmic decrement (6.5)

α_{c1} classical absorption coefficient

α_{c1}' high frequency sound absorption coefficient in gases (2.2.9)

α_d degree of dissociation (14.2)

α_{ex} excess absorption coefficient

α_0 value of logarithmic decrement when no liquid is present (6.5)

α_{obs} observed absorption coefficient

α_r ratio of τ_s at a given temperature to τ_s at a reference temperature (5.7)

α_R absorption coefficient arising from Rayleigh scattering

α_s absorption coefficient arising from shear viscosity

α_{th} absorption coefficient arising from thermal conductivity

α_{tube} absorption coefficient of sound waves in a narrow tube

β coefficient describing container and wall losses (3.4)

β $C_1\tau/C_i$ (4.5)

β parameter describing width of Davidson–Cole function (5.5.10)

β geometrical constant (6.4)

β^* complex compressibility

β_r ratio of G_∞ at a given temperature to G_∞ at a reference temperature (5.7)

β_S adiabatic compressibility

β_{Sr} relaxational part of β_S

$\beta_{S\infty}$ value of β_S at 'infinite' frequency

β_T	isothermal compressibility	λ_B	de Broglie wavelength (4.2.18)
β_{Tr}	relaxational part of β_T	λ_l	wavelength of light
$\beta_{T\infty}$	value of β_T at 'infinite' frequency	λ_p, λ_p'	eigenvalues (7.2)
γ	C_p/C_v	Λ	average number of chain bonds between entanglements (8.3)
γ	shear strain		
γ_0	value of γ ($= C_p/C_v$) at low sound frequencies	μ	absorption per wavelength $\alpha\lambda$
γ_0	initial value of shear strain	μ	parameter in potential energy function (4.2.8)
γ_ω	effective value of γ ($= C_p/C_v$) at sound frequency ω	μ'	relaxational absorption per wavelength
δ	phase angle (6.3)	μ_m	maximum absorption per wavelength
Δ	differencing symbol	μ_m'	maximum value of μ'
$\Delta G^{\neq}$	free energy of activation	ν	frequency of vibration
$\Delta H^{\neq}$	enthalpy of activation	ν_l	frequency of light
$\Delta S^{\neq}$	entropy of activation	ν_M	lowest molecular vibration frequency
$\Delta\alpha$	amplitude change (6.7)		
$\Delta\beta$	phase change (6.7)	ν_{M+1}	second lowest molecular vibration frequency
ε	depth of Lennard–Jones potential well (11.4)		
ζ	roughness parameter (4.3)	ν_p	eigenvalue
ζ	segmental friction coefficient (7.2)	ξ	amplitude of sound wave (3.9.2)
η^*	complex viscosity	ξ	displacement
η'	real part of η^*	ξ_0, ξ_0'	initial displacement
η''	imaginary part of η^*	π	3·1416
η_B	coefficient of bulk viscosity	ρ	density
η_s	coefficient of shear viscosity	ρ'	new density (2.1)
η_S	solvent viscosity	ρ_c	critical density (11.4)
η_v	coefficient of volume viscosity	ρ_Q	density of crystal quartz (6.6)
η_x	extrapolated unentangled polymer viscosity (8.1)	ρ_r	density of rod (6.7)
η_∞	viscosity of a polymer solution at 'infinite' frequency	ρ_s	density of solvent water (14.6)
θ	angle	σ	shear stress
θ	thermal expansion coefficient	σ	root-mean-square separation of ends of polymer segment (7.2)
θ_0	maximum angular displacement (6.5.3)		
θ_r	relaxational part of thermal expansion coefficient	σ	molecular diameter (11.4)
		σ_0	initial shear stress
θ_∞	value of thermal expansion coefficient at 'infinite' frequency	τ	relaxation time
		τ'	$b\tau$ (2.5.6)
		τ_B	bulk relaxation time
κ	thermal conductivity	τ_d	relaxation time of diffusion process (9.4.6)
λ	wavelength		
λ	parameter in potential energy function (4.2.8)	τ_f	relaxation time of flow process (9.4.6)

τ_p	relaxation time of pth polymer mode		ϕ	defined in (4.2.9)
			ϕ	angle (6.8.4)
τ_p'	τ_p of molecule with internal viscosity (7.2.13)		ϕ	coefficient of internal viscosity (7.2)
τ_r	rotational–translational relaxation time		ϕ	ultrasonic vibration potential (14.6)
τ_s	shear relaxation time		$\psi(t)$	stress relaxation function
τ_s'	τ at maximum of Gaussian function (5.5.8)		ω	angular frequency $2\pi f$
$(\tau_s)_j$	shear relaxation time of jth process (5.5.1)		ω_{inf}	ω at point of inflexion of velocity dispersion curve (2.5)
τ_v	volume or structural relaxation time		ω_m	ω at μ_m
ϕ	phase difference (3.9)		ω_s	relaxation frequency (5.6.9)

NOTE ON UNITS

As far as possible S.I. units have been used in this book. One exception is pressure, where the atmosphere has been retained (1 atm $= 1\cdot013 \times 10^5$ N m^{-2}). The c.g.s. equivalents of certain S.I. units are listed below.

Physical quantity	S.I. unit	c.g.s. equivalent
Temperature	K	°K
Elastic modulus	N m^{-2}	10 dyn cm^{-2}
Kinematic viscosity	m^2 s^{-1}	10^4 stokes
Dynamic viscosity	N s m^{-2}	10 poise
Mechanical impedance	N s m^{-3}	$0\cdot1$ dyn s cm^{-3}

CHAPTER 1

INTRODUCTION

1.1 Molecular Acoustics

Molecular acoustics is the study of molecules and their interactions using elastic waves of audio frequencies (< 20 kHz), ultrasonic frequencies (20 kHz–1 GHz) and hypersonic frequencies (> 1 GHz). When an elastic wave passes through a medium in which there exists an equilibrium, the effect on the equilibrium depends on the frequency of the wave. If the period of the elastic wave is much longer than the time constant (or relaxation time) for the alteration of the position of equilibrium, the latter is disturbed by the wave. If, on the other hand, the period of the wave is much shorter than the relaxation time, the wave will not 'see' the equilibrium which consequently remains undisturbed. When the period is comparable to the relaxation time, changes in the velocity of propagation and absorption coefficient of the elastic wave occur: from measurements of these changes, the relaxation time and rate constants of the equilibrium may be determined. Since the period of an elastic wave can range from 1 to 10^{-10} s, a wide range of equilibria can be studied.

There are two types of elastic waves—longitudinal and shear. Sometimes the term 'ultrasonic' is used to describe both types of wavemotion although it should strictly be confined to longitudinal waves. In a shear or transverse wave, the displacement of the particles of the medium is at right angles to the direction of propagation of the wave, while in a longitudinal wave the particles move in the direction of propagation.

When a medium experiences a small shear strain (Figure 1.1.1), there is no volume change and no temperature change. Hence a shear wave propagating in a medium does not induce a temperature variation in phase with the displacement of the medium, and any equilibria which are present are not disturbed. The medium may respond to the shear wave by viscous flow (i.e. a liquid), by elastic deformation (solid) or by some combination of the two (viscoelastic body). The change from viscous to elastic behaviour (viscoelastic relaxation) occurs when the period of the shear wave becomes comparable to the time for an elementary diffusive motion of the molecules of the medium. This time may range from many seconds in polymers to 10^{-13} s in liquid argon. The viscoelastic behaviour of polymers is discussed in Chapters 7 and 8 and that of liquids in Chapter 9.

An ultrasonic longitudinal wave contains both a pure shear and a pure compressional component (Figure 1.1.2). Since the period of the alternating compression is short, a longitudinal wave will propagate adiabatically and the local temperature of the medium will alter in phase with its changing volume. Hence any equilibria which are sensitive to temperature or pressure will be disturbed by the passage of a longitudinal wave. This

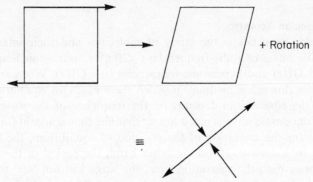

+ Rotation

FIGURE 1.1.1. Pure shear considered as the resultant of a compression and an expansion at right angles.

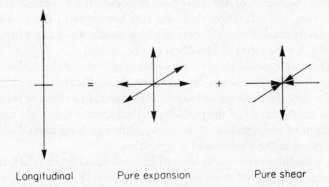

Longitudinal   Pure expansion   Pure shear

FIGURE 1.1.2. A longitudinal deformation considered as the resultant of a pure expansion and a pure shear.

includes not only chemical equilibria (Chapter 14) but also molecular energy transfer between the translational and the vibrational and rotational degrees of freedom in gases (Chapter 4) and liquids (Chapter 11), and rotational isomerization in liquids (Chapter 12). The flow of liquid molecules between packings of high and low density can also be studied with longitudinal waves (Chapter 10).

The experimental techniques used to study the propagation of elastic waves in gases, liquids and solids are discussed in Chapters 3 and 6. It is generally true that longitudinal waves propagate more easily through a given medium than do shear waves. Hence longitudinal waves are studied by propagation methods and their velocity and absorption coefficient determined. Shear wave techniques rely mainly on the reflexion of a wave at an interface between the generating transducer and the medium to be investigated.

A mathematical treatment of the propagation of elastic waves in a medium is given in Chapters 2 and 5. A rigorous solution of the equations describing wavemotion in a relaxing medium has not yet been obtained. The treatment given here is adequate for the derivation of most molecular parameters, and some of the approximations involved have been indicated.

We shall consider the propagation of elastic waves of low amplitude only. The medium under investigation is gently perturbed by such waves, but no permanent changes are induced. We shall confine our attention mainly to gases, liquids and polymers since the phenomena observed there can be readily interpreted on a molecular basis. Nevertheless much effort has been devoted to studying the properties of solids by acoustic techniques, and a brief survey of some of the results of this work is given in Chapter 15.

1.2 Historical Background

The 'classical' absorption of sound in gases was first investigated theoretically by Stokes (1845) and Kirchhoff (1868) and found to increase with the square of the sound frequency. The experimental work of Duff (1898), Altberg (1907) and Neklepajew (1911) showed that the observed absorption was substantially greater than the classical value. With the advent of thermionic valves, Pierce (1925) developed the ultrasonic interferometer and observed the high sound absorption and velocity dispersion in gaseous CO_2. The possibility of a finite time delay in transferring energy from translation to vibration and rotation in gases was suggested independently by Lorentz (1881), Boltzmann (1895) and Rayleigh (1899). In 1920 Einstein showed that measurements of sound velocity could be used to determine the rate of dissociation of gaseous N_2O_4, while Herzfeld and Rice (1928) first developed a relation between the sound velocity and excess absorption and the rate of a reaction.

The absorption coefficient of an ultrasonic longitudinal wave in a liquid is typically one hundred times less than that in a gas at the same frequency and can be measured less readily. Although Hubbard and Loomis (1927) adapted the ultrasonic interferometer to measure ultrasonic velocity in liquids, the first measurement of absorption in liquids was by Biquard (1931). Biquard developed a pulse technique for measuring velocity in liquids (Biquard and Ahier, 1943) and also an optical method for sound absorption (Biquard, 1933). The study of liquids was greatly facilitated by the development from radar principles of the pulse technique by Pellam and Galt (1946) and Pinkerton (1947). The first theory of ultrasonic absorption in liquids was proposed by Kneser (1938): he suggested that the excess absorption arose from the same mechanism as in gases, namely molecular energy transfer from translation to the internal modes. Pinkerton (1949) concluded that a number of different processes can lead to excess absorption in liquids and derived a useful classification of liquids in terms of their ultrasonic properties.

The viscosity and elasticity of matter have been studied since the earliest times, and Lucretius made several rheological observations. Many devices were used to study viscoelastic behaviour under steady shear and these developed into low frequency vibrating systems. A major advance was the application of piezoelectric methods of generating shear waves to the study of viscoelasticity (Mason, 1947; Mason and coworkers, 1949; McSkimin, 1952). Molecular theories of viscoelasticity have developed mainly from the statistical theories of polymer viscosity and rubber elasticity (Guth and Mark, 1934; Kuhn, 1934).

A review of the development of molecular acoustics has been given by Herzfeld (1966), while Bergmann (1949) also gives an account of earlier work. Useful introductory texts dealing with several aspects of the subject are those of Vigoureux (1950), Richardson (1962), Schaaffs (1963), Blitz (1967) and Gooberman (1968). Nozdrev (1965) places particular emphasis on Russian work, while Cottrell and McCoubrey (1961), Ferry (1961) and Mason (1958) discuss gases, polymers and solids respectively. Two comprehensive books are those of Herzfeld and Litovitz (1959) and Bhatia (1967). Finally the series of volumes edited by Mason (1964–1970) covers all aspects of physical and molecular acoustics in considerable detail.

CHAPTER 2

SOUND PROPAGATION

2.1 Velocity of Plane Waves in a Medium

Most studies in molecular acoustics are carried out using plane waves of low amplitude. We consider first the propagation of an ultrasonic longitudinal wave along a wide uniform tube of unit cross-section. This tube contains an isotropic homogeneous medium which may be either a fluid or a solid. The particles of the medium move in an oscillatory manner in the direction of wave propagation as a result of the alternating compressions and rarefactions caused by the sound wave. We consider the motion of an element of the medium which is small in comparison with the sound wavelength and which is initially contained between two parallel plates A and B (Figure 2.1.1). These plates are at right angles to the

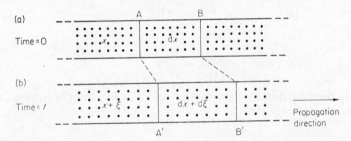

FIGURE 2.1.1. Propagation of a longitudinal ultrasonic wave along a wide tube: (a) section through length of tube at time zero; (b) section through length of tube at time t during a rarefaction.

direction of propagation of the sound wave and are at distances x and $(x+dx)$ from some reference point in the tube. ρ is the initial density of the medium, and the mass of the element is $\rho\,dx$. After a time t the element is displaced to a new position A'B' in the tube where A' is a distance $(x+\xi)$ from the origin and the length of the element is now $(dx+d\xi)$. Since the mass of the element remains constant, the new value of the density is

$$\rho' = \rho\,\frac{dx}{dx+d\xi} \approx \rho\left(1-\frac{\partial\xi}{\partial x}\right) \tag{2.1.1}$$

for small displacements. The net force acting on the displaced element is $(P_{A'} - P_{B'})$ where P is the pressure, and in a loss-free system this equals the product of the mass of the element and its acceleration:

$$\rho \, dx \, \frac{d^2 \xi}{dt^2} = P_{A'} - P_{B'} = -\frac{\partial P}{\partial x} \, dx \qquad (2.1.2)$$

Hence

$$\rho \frac{d^2 \xi}{\partial t^2} = -\frac{\partial P}{\partial x}$$

$$= -\frac{\partial P}{\partial \rho} \frac{\partial \rho}{\partial x}$$

$$= \rho \frac{\partial P}{\partial \rho} \frac{d^2 \xi}{dx^2} \quad \text{from equation (2.1.1).} \qquad (2.1.3)$$

The equation of motion of the element of the medium is thus

$$\frac{d^2 \xi}{dt^2} = \frac{\partial P}{\partial \rho} \frac{d^2 \xi}{dx^2}$$

or

$$\frac{d^2 \xi}{dx^2} = \frac{d^2 \xi}{dt^2} \Big/ \frac{\partial P}{\partial \rho} \qquad (2.1.4)$$

This is the equation of wave motion. $(\partial P / \partial \rho)$ has the dimensions of the square of a velocity U:

$$U^2 = \frac{\partial P}{\partial \rho} \qquad (2.1.5)$$

The solution of equation (2.1.4) for the usual case in which the particle displacement, velocity, pressure, etc. vary sinusoidally with time is

$$\xi = \xi_0 \exp i\omega(t - x/U) + \xi_0' \exp i\omega(t + x/U) \qquad (2.1.6)$$

Here ξ_0 and ξ_0' are constants, ω is the angular frequency and U is the phase velocity. The motion of the medium can be represented by two waves travelling in opposite directions with velocity U. The second term in (2.1.6) refers to a wave reflected at some obstacle in the tube: if no reflexion occurs, $\xi_0' = 0$, and there is only a single progressive wave of maximum amplitude ξ_0. Equation (2.1.6) is valid only if $d\xi/dx$ in equation (2.1.1) is small compared to unity, that is if the amplitude of the sound wave is small. We have also assumed that there is no attenuation of the sound wave in the medium.

Equation (2.1.5) may be solved provided that some assumption is made as to the conditions which govern the pressure change. If we suppose that the processes in a sound wave are adiabatic and reversible,

$$U^2 = (\partial P/\partial \rho)_S \tag{2.1.7}$$

For an ideal gas under adiabatic conditions

$$PV^\gamma = \text{constant} \tag{2.1.8}$$

where V is the molar volume of the gas, and γ is the ratio of the molar heat capacities. Then

$$U^2 = \frac{\gamma P}{\rho} = \frac{\gamma RT}{M} \tag{2.1.9}$$

R being the gas constant, T the absolute temperature and M the molecular weight of the gas. Equation (2.1.9) is applicable only to an ideal gas: a correction is required to convert the observed ultrasonic velocity in a real gas to the value for the ideal gas (Cottrell and McCoubrey, 1961). Conversely, the sound velocity in a real gas may be used to calculate the second virial coefficient of the gas (Cottrell, Macfarlane and Read, 1965; Lestz and Grove, 1965).

In a liquid where $V = 1/\rho$, we have

$$\left(\frac{\partial P}{\partial \rho}\right)_S = -1 \Big/ \frac{1}{V^2}\left(\frac{\partial V}{\partial P}\right)_S = 1/\rho\beta_S = \gamma/\rho\beta_T \tag{2.1.10}$$

Here

$$\beta_S = -\frac{1}{V}\left(\frac{\partial V}{\partial P}\right)_S \quad \text{is the adiabatic compressibility,}$$

$$\beta_T = -\frac{1}{V}\left(\frac{\partial V}{\partial P}\right)_T \quad \text{is the isothermal compressibility,}$$

and

$$\beta_T/\beta_S = \gamma.$$

Hence

$$U^2 = \frac{1}{\rho\beta_S} = \frac{\gamma}{\rho\beta_T} \tag{2.1.11}$$

In an isotropic solid it is convenient to replace the compressibility by the longitudinal elastic modulus M, giving

$$U^2 = \frac{M_S}{\rho} = \frac{\gamma M_T}{\rho} \tag{2.1.12}$$

The above treatment refers to longitudinal ultrasonic waves in which the particle displacement in the medium is in the direction of propagation of the wave. Ultrasonic transverse or shear waves are also of considerable importance in molecular acoustics: here the particle displacement in the medium is at right angles to the direction of propagation of the wave. Shear waves are strongly absorbed in gases and liquids: their motion cannot be described by a simple treatment analogous to that given above and is discussed in Section 5.4. The absorption of shear waves in solids is generally small, however, and the velocity of shear waves U_s in a solid is given by an equation similar to equation (2.1.12)

$$U_s^2 = G/\rho \qquad (2.1.13)$$

where G is the rigidity modulus of the solid.

2.2   Absorption of Plane Longitudinal Waves in Gases and Low Viscosity Liquids when Relaxational Effects are Absent

An ultrasonic longitudinal wave is attenuated in passing through a real medium. If the medium has a finite viscosity, frictional losses occur and Rayleigh (1877) showed that the pressure P in equation (2.1.2) must be replaced by an effective pressure

$$P - \frac{4}{3}\eta_s \frac{\partial}{\partial t}\left(\frac{\partial \xi}{\partial x}\right) \qquad (2.2.1)$$

η_s is the coefficient of shear viscosity defined as the ratio of the shear stress in a plane to the particle velocity gradient along the normal to the plane. The equation of motion (2.1.4) then becomes

$$\frac{\partial^2 \xi}{\partial t^2} = U^2 \frac{\partial^2 \xi}{\partial x^2} + \frac{4}{3}\frac{\eta_s}{\rho}\frac{\partial^3 \xi}{\partial x^2 \partial t} \qquad (2.2.2)$$

If the absorption is small, the sound velocity U has almost the same value as in a non-dissipative medium. The solution of this equation for a sinusoidal wave is

$$\xi = \xi_0 \exp(-\alpha x)\exp[i\omega(t - x/U)] \qquad (2.2.3)$$

α is the absorption coefficient of the sound wave in the medium, and the ratio I_1/I_2 of the sound intensities at two points a distance x apart in the direction of propagation is 1 to $\exp(-2\alpha x)$: the amplitude of the sound wave decreases by a factor of $\exp(-\alpha x)$ in the same distance. The units of α are nepers per metre, although α is often expressed in decibels per metre. Since the ratio of two intensities is $10\log_{10}(I_1/I_2)$ dB, one neper equals 8·686 dB.

The value of α_s for shear viscosity may be obtained by inserting (2.2.3) into (2.2.2) to give

$$\alpha_s = \frac{8}{3}\frac{\eta_s}{\rho}\frac{\pi^2 f^2}{U^3} \tag{2.2.4}$$

where $f = \omega/2\pi$ is the frequency.

A second factor which leads to absorption of longitudinal waves is the thermal conductivity of the medium. At any instant the high pressure regions will have a temperature above the average while the temperature of the low pressure regions in the medium will be below the average. Heat will be conducted from the high to the low temperature regions, and a compressed region will return less work on expansion than was required to compress it. This leads to a sound absorption α_{th}, and it can be shown (Herzfeld and Litovitz, 1959) that

$$\alpha_{th} = \frac{2\pi^2 \kappa(\gamma-1)}{\rho\gamma C_v U^3} f^2 \tag{2.2.5}$$

where κ is the thermal conductivity and C_v is the heat capacity at constant volume.

The sum of the losses caused by viscosity and thermal conductivity is called the classical absorption α_{cl}:

$$\alpha_{cl} = \alpha_s + \alpha_{th} = \frac{2\pi^2}{\rho U^3}\left(\frac{4}{3}\eta_s + \frac{\gamma-1}{\gamma}\frac{\kappa}{C_v}\right)f^2 \tag{2.2.6}$$

The classical absorption of ultrasonic longitudinal waves in a medium increases as the square of the ultrasonic frequency. Except in monatomic gases the experimentally observed absorption α_{obs} is always larger than this classical value. It is customary to attribute this excess absorption α_{ex} to a bulk viscosity η_B, and (2.2.1) is replaced by

$$P - \left(\frac{4}{3}\eta_s + \eta_B\right)\frac{\partial}{\partial t}\left(\frac{\partial \xi}{\partial x}\right) \tag{2.2.7}$$

Hence

$$\alpha_{ex} = \alpha_{obs} - \alpha_{cl} = \frac{2\pi^2 f^2}{\rho U^3}\eta_B \tag{2.2.8}$$

The bulk viscosity of a fluid arises from two sources. The first is the energy loss associated with the flow of liquid molecules between positions of different density during a change in the volume of a given mass of liquid (Chapter 10). We shall call this the volume viscosity η_v. The other contribution to the bulk viscosity comes from relaxation processes such as rotational isomerization (Chapter 12) or vibrational energy transfer

(Chapter 11). A number of different terms are used in the literature to describe these viscosities, for example structural viscosity or compressional viscosity. We shall use the term bulk viscosity as defined in equation (2.2.7) to include all causes of sound absorption other than shear viscosity and thermal conductivity, while the term volume viscosity will be restricted to the contribution of structural processes to the bulk viscosity (Chapter 10). This bulk viscosity should not be confused with the so-called bulk viscosity of polymers which refers to the steady flow shear viscosity of the bulk, undiluted polymer.

At frequencies above 10^8 Hz in gases at atmospheric pressure, the sound frequency becomes comparable to the time between collisions. This leads to a higher sound absorption than is predicted by equation (2.2.6) which is derived on the assumption that the gas behaves as a continuum. Greenspan (1954, 1956) has derived an approximate expression for the ultrasonic absorption in this region α'_{cl}

$$\alpha'_{cl} = \frac{2\pi}{U}\left[\frac{-1+(1+z^2)^{\frac{1}{2}}}{2(1+z^2)}\right]f \qquad (2.2.9)$$

where $z = 2 \cdot 6\pi f \eta_s / P$. Another approximation, that of Burnett (1935), gives good agreement with the observed absorption in monatomic gases. This increased absorption is accompanied by velocity dispersion (Greenspan, 1965).

In a fluid mixture there is an additional mechanism of ultrasonic absorption. The pressure and temperature variations in a longitudinal sound wave produce local variations in the concentration of the mixture. These concentration inequalities are reduced by diffusion, and this leads to attenuation of the sound wave. Kohler (1941, 1950) has made a kinetic theory calculation of the effect of diffusion on ultrasonic absorption in ideal gas mixtures, and this theory has been verified experimentally by Law, Koronaios and Lindsay (1967).

2.3 General Considerations on Relaxation

Consider a two-state equilibrium

$$A \underset{k_b}{\overset{k_f}{\rightleftharpoons}} B$$

where k_f and k_b are the forwards and backwards rate constants. We assume that the energy of B is greater than that of A: A and B may be different chemical species, or B may be some excited state of A. Let n_A and n_B be the actual number of particles of A and B at any instant, and let the total number of particles be

$$n_A + n_B = N \qquad (2.3.1)$$

The rate of disappearance of B species at any instant is

$$-\frac{dn_B}{dt} = k_b n_B - k_f n_A$$

$$= (k_b + k_f) n_B - k_f N \quad \text{using (2.3.1)}$$

$$= (k_b + k_f) \left[n_B - \left(\frac{1}{(1 + k_b/k_f)} \right) N \right] \tag{2.3.2}$$

At equilibrium, $dn_B/dt = 0$, and

$$k_b n_B^e = k_f n_A^e \tag{2.3.3}$$

the superscript e indicating the equilibrium number of A and B species. Substituting the value of k_b/k_f from (2.3.3) into (2.3.2), we obtain

$$-\frac{dn_B}{dt} = (k_b + k_f)(n_B - n_B^e)$$

$$= (n_B - n_B^e)/\tau \tag{2.3.4}$$

τ is defined by equation (2.3.4) as $1/(k_b + k_f)$, and is the relaxation time of the equilibrium.

The significance of τ may be understood by integrating equation (2.3.4). We have

$$\frac{dn_B}{n_B^e - n_B} = \frac{dt}{\tau} \tag{2.3.5}$$

Hence

$$\ln \left(\frac{1}{n_B^e - n_B} \right) = \frac{t}{\tau} + \text{constant} \tag{2.3.6}$$

The value of the constant of integration may be obtained by noting that at $t = 0$, n_B equals its initial value n_B^0. Hence

$$\frac{t}{\tau} = \ln \left(\frac{n_B^e - n_B^0}{n_B^e - n_B} \right) \tag{2.3.7}$$

that is

$$\frac{n_B - n_B^e}{n_B^0 - n_B^e} = e^{-t/\tau} \tag{2.3.8}$$

This equation is shown graphically in Figure 2.3.1. The distance of the reaction from equilibrium falls exponentially with time, τ being a measure of the rate of this approach to equilibrium. After a time equal to the relaxation time τ, the distance from equilibrium has fallen from its initial value of $(n_B^0 - n_B^e)$ to $1/e$ of this value.

In many cases of interest $k_b \gg k_f$. If this is so, $\tau \approx 1/k_b$, and, since $n_B^e \approx 0$, equation (2.3.8) reduces to

$$n_B/n_B^0 \approx e^{-t/\tau} \qquad (2.3.9)$$

Such a situation sometimes occurs when B is an excited form of A such as a vibrationally excited state. The relaxation time τ is then a direct measure of the lifetime of the excited state.

The discussion in this section refers only to a two-state equilibrium. Often more than two states are present as in the case of the harmonic

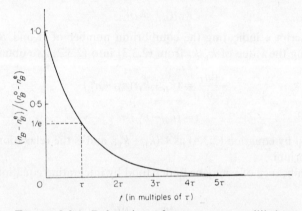

FIGURE 2.3.1. Relaxation of a two-state equilibrium $A \rightleftharpoons B$ with an initial number of B particles n_B^0 and a final equilibrium number n_B^e. After a time τ, the displacement from equilibrium has fallen to $1/e$ of its initial value.

oscillator or when more than two chemical species take part in the equilibrium. The expressions for τ in such cases will differ from those of the two-state case (Herzfeld and Litovitz, 1959; Czerlinski, 1966).

2.4 Effective Heat Capacity of a Relaxing Medium

We shall now consider the effect of propagating a sinusoidal longitudinal wave through the two-state system discussed in the previous Section. On account of the temperature variations in the sound wave, the equilibrium population of the two states will change. If the frequency of the sound wave is sufficiently low, the local value of n_B^e will remain in equilibrium with the local temperature in the sound wave. For a sinusoidal wave we can write

$$n_B^e = A' \exp(i\omega t) \qquad (2.4.1)$$

where A' is a constant.

As the frequency of the sound wave increases, the local value of n_B (which we shall call n_B^ω) may be unable to follow the rapid temperature changes if k_b or k_f is too small. n_B^ω will relax towards the instantaneous equilibrium value n_B^e according to equation (2.3.4):

$$-\frac{dn_B^\omega}{dt} = \frac{(n_B^\omega - n_B^e)}{\tau} \qquad (2.4.2)$$

This equation can be rewritten in a more convenient form by noting that for a sinusoidal variation such as that in equation (2.4.1), a time derivative d/dt may be replaced by $i\omega$ since

$$\frac{dn_B^e}{dt} = i\omega A' \exp(i\omega t) = i\omega n_B^e \qquad (2.4.3)$$

Hence

$$-i\omega n_B^\omega = (n_B^\omega - n_B^e)/\tau$$

or

$$n_B^\omega = \frac{n_B^e}{1 + i\omega\tau} = \frac{A' \exp(i\omega t)}{1 + i\omega\tau} \qquad (2.4.4)$$

We now write the total heat capacity at constant volume of the system C_v^ω as the sum of two components: one, C_i^ω, represents the contribution to C_v^ω of the equilibrium $A \rightleftharpoons B$, while the other, $C_{v\infty}$, incorporates all the other contributions of A and B. At low sound frequencies where the position of equilibrium is exactly in phase with the sound wave, the effective relaxational heat capacity C_i^ω will equal C_i. At high sound frequencies where the period of the sound wave is too short to promote A molecules to state B, the effective relaxational heat capacity will be zero. At intermediate frequencies C_i^ω will depend on the number of A molecules which the sound wave is able to convert to state B during the period of the wave, so that

$$\frac{C_i^\omega}{C_i} = \frac{n_B^\omega}{n_B^e} = \frac{1}{1 + i\omega\tau} \qquad (2.4.5)$$

by (2.4.4). The total heat capacity at constant volume of the two-state system is therefore

$$C_v^\omega = C_{v\infty} + \frac{C_i}{1 + i\omega\tau} \qquad (2.4.6)$$

At low sound frequencies, where $\omega\tau \ll 1$

$$C_v^\omega = C_{v\infty} + C_i = C_v \qquad (2.4.7)$$

while at high sound frequencies $(\omega\tau \gg 1)$

$$C_v^\omega = C_{v\infty} \qquad (2.4.8)$$

2.5 Propagation of Ultrasonic Longitudinal Waves in a Relaxing Medium

The effect of a relaxation process is to introduce an extra absorption of the sound wave: the velocity of propagation of the sound wave is also altered slightly. The wave equation (2.2.3) may be written

$$\xi = \xi_0 \exp\left[i\omega\left\{t - x\left(\frac{1}{U} - \frac{i\alpha}{\omega}\right)\right\}\right] \tag{2.5.1}$$

Here α is the extra ultrasonic absorption arising from the relaxation process: we assume that this is additive with the absorption from shear viscosity and thermal conductivity (Section 2.2). The velocity U^* of such a sound wave is complex and is given by

$$\frac{1}{U^*} = \frac{1}{U} - \frac{i\alpha}{\omega} \tag{2.5.2}$$

This equation may be rewritten in terms of the 'low frequency' velocity U_0 using equation (2.1.11)

$$\frac{U_0^2}{U^{*2}} = \left(\frac{U_0}{U} - \frac{i\alpha U_0}{\omega}\right)^2 = \frac{\gamma}{\gamma_\omega} \tag{2.5.3}$$

γ is the ratio of the molar heat capacities of the medium at low sound frequencies while γ_ω is the effective heat capacity ratio for a sound wave of frequency ω. In deriving equation (2.5.3), we have assumed that the isothermal compressibility β_T in (2.1.11) is independent of the relaxation process. This is equivalent to assuming that the equation of state of the fluid depends only on the translational temperature (Herzfeld and Litovitz, 1959). This assumption is not always valid in liquids, and in such cases the treatments in Chapters 10 and 12 must be used.

The value of γ_ω in (2.5.3) may be calculated from the value of C_v^ω given by (2.4.6) and the analogous expression for C_p^ω. Hence

$$\left(\frac{U_0}{U} - \frac{i\alpha U_0}{\omega}\right)^2 = \frac{C_p}{C_v}\left(C_{v\infty} + \frac{C_i}{1 + i\omega\tau}\right)\left(C_{p\infty} + \frac{C_i}{1 + i\omega\tau}\right)^{-1}$$

$$= \left(1 + \frac{C_{v\infty}i\omega\tau}{C_v}\right)\left(1 + \frac{C_{p\infty}i\omega\tau}{C_p}\right)^{-1} \tag{2.5.4}$$

In order to simplify the calculation of the ultrasonic velocity and absorption coefficient from this equation, we use the relation (cf. Herzfeld and Litovitz, 1959)

$$\frac{1 + ai\omega\tau}{1 + bi\omega\tau} = 1 + (a - b)\frac{i\omega\tau}{1 + bi\omega\tau} = 1 + \frac{a - b}{b}\frac{i\omega\tau'}{1 + i\omega\tau'} \tag{2.5.5}$$

where

$$\tau' = b\tau \tag{2.5.6}$$

From equation (2.5.4) we have

$$a = C_{v\infty}/C_v = (C_v - C_i)/C_v \tag{2.5.7}$$

$$b = C_{p\infty}/C_p = (C_p - C_i)/C_p \tag{2.5.8}$$

and

$$a - b = - C_i R'/C_v C_p \tag{2.5.9}$$

Here R' is the heat capacity difference $(C_p - C_v)$: in an ideal gas, $R' = R$, the gas constant.

Equation (2.5.4) may now be rewritten

$$\left(\frac{U_0}{U} - \frac{i\alpha U_0}{\omega}\right)^2 = 1 - \frac{R' C_i}{C_v(C_p - C_i)} \frac{i\omega\tau'}{1 + i\omega\tau'}$$

$$= 1 - \frac{R' C_i \omega^2 \tau'^2 + R' C_i i\omega\tau'}{C_v(C_p - C_i)(1 + \omega^2 \tau'^2)} \tag{2.5.10}$$

In order to calculate U and α, we equate the real and imaginary parts of equation (2.5.10). This gives

$$\frac{U_0^2}{U^2} - \frac{\alpha^2 U_0^2}{\omega^2} = 1 - \frac{R' C_i}{C_v(C_p - C_i)} \frac{\omega^2 \tau'^2}{1 + \omega^2 \tau'^2} \tag{2.5.11}$$

and

$$\frac{2\alpha U_0^2}{U\omega} = \frac{R' C_i}{C_v(C_p - C_i)} \frac{\omega\tau'}{1 + \omega^2 \tau'^2} \tag{2.5.12}$$

Herzfeld and Litovitz (1959) have shown that in many cases the absorption of sound in a medium is so small that

$$\frac{\alpha U_0^2}{\omega^2} \ll \frac{U_0^2}{U^2} \tag{2.5.13}$$

In this case, equation (2.5.11) reduces to

$$\frac{U_0^2}{U^2} = 1 - \frac{R' C_i}{C_v(C_p - C_i)} \frac{\omega^2 \tau'^2}{1 + \omega^2 \tau'^2} \tag{2.5.14}$$

Equation (2.5.14) gives the ratio of the low frequency sound velocity U_0 to the actual sound velocity U at frequency ω in terms of the thermodynamic properties of the medium and the relaxation time of the equilibrium which is being perturbed. At sound frequencies which are low compared with the inverse of the relaxation time ($\omega\tau' \ll 1$), (2.5.14) reduces to $U_0^2/U^2 = 1$. At high sound frequencies ($\omega\tau' \gg 1$), $U = U_\infty$, and

$$\frac{U_0^2}{U_\infty^2} = 1 - \frac{R' C_i}{C_v(C_p - C_i)} \tag{2.5.15}$$

Equation (2.5.14) may be rearranged with the help of (2.5.15) to give

$$\omega^2 \tau'^2 = \frac{U_\infty^2(U^2 - U_0^2)}{U_0^2(U_\infty^2 - U^2)} \tag{2.5.16}$$

The variation of U^2 with $\log \omega$ is given in Figure 2.5.1. For an increase in U^2 from 0.1 $(U_\infty^2 - U_0^2)$ to 0.9 $(U_\infty^2 - U_0^2)$, ω increases by just under a factor of ten. The curve has a point of inflexion at

$$\omega_{\text{int}} = \frac{1}{\tau'} \frac{U_\infty}{U_0} \tag{2.5.17}$$

and thus τ' may be evaluated. τ is then obtained from equations (2.5.6) and (2.5.8).

The frequency dependence of the ultrasonic absorption is conventionally discussed in terms of absorption per wavelength $\mu (= \alpha\lambda)$ or of α/f^2. From (2.5.12) we have

$$\mu = \frac{U^2}{U_0^2} \frac{\pi R' C_i}{C_v(C_p - C_i)} \frac{\omega\tau'}{1 + \omega^2 \tau'^2} \tag{2.5.18}$$

If the velocity dispersion is small, $U \approx U_0$, and (2.5.18) reduces to

$$\mu \approx \frac{\pi R' C_i}{C_v(C_p - C_i)} \frac{\omega\tau'}{1 + \omega^2 \tau'^2} \tag{2.5.19}$$

Figure 2.5.2 shows the bell-shaped curve given by (2.5.19). The maximum value μ_m occurs at ω_m where $\omega_m \tau' = 1$, while

$$\mu_m = \frac{\pi R' C_i}{2 C_v(C_p - C_i)} \tag{2.5.20}$$

Equation (2.5.12) can also be rearranged to give

$$\frac{\alpha}{f^2} \approx \frac{2\pi^2}{U_0} \frac{R' C_i}{C_v(C_p - C_i)} \frac{\tau'}{1 + \omega^2 \tau'^2} \tag{2.5.21}$$

This is often written

$$\frac{\alpha}{f^2} = \frac{A}{1 + \omega^2 \tau'^2} + B \tag{2.5.22}$$

where A and B are constants, B representing the absorption from processes other than the relaxation. As seen in Figure 2.5.3, α/f^2 falls from a constant low frequency value A to another constant value B at high frequencies. The point of inflexion again occurs when $\omega_m \tau' = 1$.

The preceding equations may be evaluated particularly easily when sound propagation in an ideal gas is being considered. The contribution of the translational degrees of freedom to C_v is then $1.5R$, while that of the rotational degrees of freedom is zero, R or $1.5R$ depending on whether the

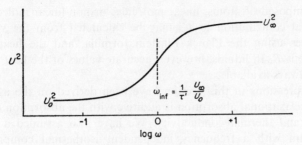

FIGURE 2.5.1.

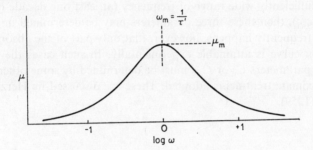

FIGURE 2.5.2.

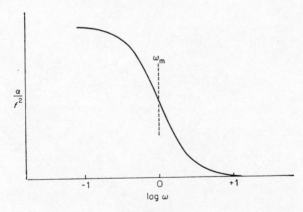

FIGURE 2.5.3.

FIGURES 2.5.1 to 2.5.3. Square of the ultrasonic velocity U^2, absorption per wavelength μ, and absorption per metre divided by the square of the ultrasonic frequency α/f^2 plotted against the logarithm of the angular frequency ω for a single relaxation process: ω is in arbitrary units.

gas is composed of atoms, linear molecules or non-linear molecules. The vibrational contribution to C_v may be calculated from the vibrational frequencies using the Planck–Einstein formula, and the heat capacity difference is R. In liquids, however, accurate values of the heat capacities are not always available.

The expressions in this Section have been derived on the assumption that the relaxational absorption is additive with the absorption caused by viscosity and thermal conductivity. We have also assumed a two-state equilibrium with a frequency independent isothermal compressibility. The equations describe the relaxation process by three parameters: τ, C_v and $C_{v\infty}$. If measurements of absorption and velocity dispersion cover a sufficiently wide range of frequency (at least one decade on either side of ω_{inf}), then these three parameters may be determined unambiguously. It frequently happens, however, that only part of the absorption or dispersion curve is attainable experimentally. In such cases the thermodynamic parameters C_v or $C_{v\infty}$ must be determined by some other means, or approximate treatments adopted. These are discussed by Herzfeld and Litovitz (1959).

CHAPTER 3

EXPERIMENTAL METHODS OF STUDYING ULTRASONIC PROPAGATION

3.1 The Generation and Detection of Ultrasonic Waves

In Chapter 2, the relations between the velocity and absorption coefficient of ultrasonic waves and the relaxation time of an equilibrium were derived. In this Chapter we consider experimental methods for studying the propagation of ultrasonic longitudinal waves of various frequencies in different media.

The most widely used transducers for converting electrical energy to acoustical energy and vice versa are based on the piezoelectric effect (Cady, 1946; Mason, 1950). A solid is said to be piezoelectric if electric charges are produced on its surface when it is subjected to a mechanical stress. If the direction of the stress is reversed, so is the polarity of the electric charge. Provided that the solid is not strained beyond its elastic limit, the charge density is directly proportional to the applied stress. Piezoelectric substances also exhibit the converse piezoelectric effect: the substance changes in size when an electric field is applied to it, the sign of this change being reversed when the direction of the field is reversed.

These effects were shown to exist in crystals which lack a centre of symmetry by Curie and Curie (1880). One of the most widely used piezoelectric crystals is quartz on account of its mechanical and chemical stability. Quartz belongs to the trigonal crystallographic system, and a simplified section through part of a quartz crystal is shown in Figure 3.1.1. The Z optic axis runs along the crystal, the crystal section at right angles to the Z axis being hexagonal in shape. The three axes joining opposite corners of this hexagon are known as the X axes, while the three axes joining opposite faces are called the Y axes.

The disposition of the silicon and oxygen atoms in the crystal lattice is such that plates cut from the crystal with their faces normal to an X or Y axis display the piezoelectric effects. X-cut crystals with faces normal to the X axis are used in the generation and detection of ultrasonic longitudinal waves. Suppose that an X-cut plate like that in Figure 3.1.1 is silvered on opposite faces and that an alternating voltage of frequency f is applied across these electrodes. For a given instantaneous electric field direction across the crystal, the latter will be compressed in the X direction and expanded in the normal Y direction. When the electric field is reversed, expansion will occur along the X axis and contraction along the Y. Thus the crystal dimensions will oscillate at the frequency f. These oscillations will be of small amplitude unless f coincides with one of the natural

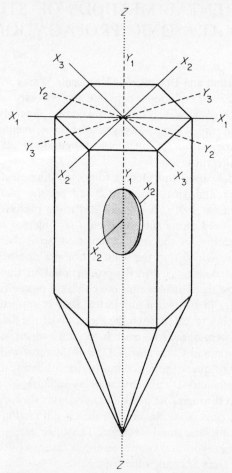

FIGURE 3.1.1. Simplified diagram of part of a quartz crystal, showing the Z, Y and X axes. The cylindrical disc represents an X-cut transducer.

frequencies of mechanical vibration of the quartz disc: the amplitude of the oscillation will then increase many hundreds of times, being limited only by the internal friction in the quartz and the damping of the medium surrounding the disc. The surface of the disc will ideally vibrate up and down as a plane piston to produce a plane sound wave in the surrounding medium. Conversely, a plane sound wave arriving at the crystal will induce an alternating voltage across the crystal electrodes.

The fundamental resonance frequency of a quartz plate is inversely proportional to its thickness, being 1 MHz for an X-cut crystal 2·8 mm thick. For frequencies above about 10 MHz, crystals vibrating in their fundamental mode become so thin and fragile that it is preferable to operate at a harmonic of a crystal with a lower fundamental resonance frequency. Although the acoustic energy produced decreases at high harmonics, sound frequencies up to 10 GHz have been generated in this way. At frequencies above 100 MHz, however, it is often preferable to excite non-resonant vibrations of a piezoelectric crystal. This is achieved by surface excitation (Baranskii, 1957) in which two electrodes are placed on one surface of the crystal, and an intense alternating electric field applied between them: the penetration of this field into the crystal produces mechanical vibrations in the crystal which can be transmitted through the crystal and into the surrounding medium.

The generation of ultrasonic waves by surface excitation of a piezoelectric crystal or by driving a crystal at a high harmonic is inefficient. Moreover the bonding of such a crystal to a delay line (Section 3.7) or to a solid specimen presents considerable problems. These difficulties have been overcome by the development of vapour-deposited, thin-film, insulating, piezoelectric transducers. The surface of the solid substrate is first coated with a thin conducting metal film. Then cadmium and sulphur, for example, are evaporated simultaneously from separately controlled crucibles. By adjusting the vapour pressures of the cadmium and sulphur, a highly oriented film may be obtained of any desired thickness (deKlerk and Kelly, 1965). The transducer is completed by evaporating a further metallic film. Ultrasonic frequencies from 10 MHz to 100 GHz have been obtained from CdS transducers, while ZnS and AlN have also been used (deKlerk and Kelly, 1965; Wauk and Winslow, 1968; Curtis, 1969; Llewellyn, Montagu-Pollock and Dobbs, 1969).

Although crystal quartz has many advantages over other ultrasonic transducers, it cannot be used at temperatures above its Curie temperature of 846 K at which it ceases to be piezoelectric. For higher temperature work it is necessary to generate the sound waves outside the high temperature zone and then introduce them to the hot region. Many other substances show piezoelectric properties, and ceramic transducers such as barium titanate, lead niobate, or lead zirconate containing some 10 mol% of lead titanate are sometimes used. These transducers have high internal damping and are particularly suitable for the generation of very short ultrasonic pulses. They are also useful for producing sound waves of high intensity.

At audio frequencies the dimensions of piezoelectric transducers would be prohibitively large. Various types of loudspeakers and microphones may then be used, including the ribbon loudspeaker and the capacitance microphone. Details of these are given in books on acoustics such as Kinsler and Frey (1962) and Stephens and Bate (1966). Electrostatic transducers have proved useful for studying the propagation of sound in gases in the frequency range 10 kHz–1 MHz. They consist of a thin sheet of plastic stretched tightly over a roughened metal surface. The outer side of the plastic is thinly metallized so that a d.c. voltage of some 100 V can be applied across the plastic. This potential difference leads to the attraction of the metallized plastic towards the rigid metal surface. If an alternating voltage of a few volts is superimposed on the d.c. field, the attractive force will vary sinusoidally at this frequency so that the plastic sheet will vibrate and produce a sound wave in the surrounding gas. Such transducers have the advantage of being non-resonant and thus they can be driven over a range of frequencies. Moreover their diameter can be 0·1 m or more, which is greater than is practicable with quartz crystals. Such electrostatic transducers with a plastic dielectric are sometimes called 'Sell' transducers after Sell (1937): problems in their design are discussed by Kuhl, Schodder and Schröder (1954), Meyer and Sessler (1957) and Meyer (1960).

The interpretation of the observed velocity and absorption coefficient of sound waves in different media is greatly simplified if measurements are made with plane sound waves. When the dimensions of a transducer are small compared with the sound wavelength, the transducer behaves essentially as a point source emitting acoustic energy in all directions, the wavefronts being spherical. To obtain an approximately plane wavefront, the dimensions of the transducer should be much greater than the sound wavelength in the medium to be investigated. The exact shape of the resultant wavefront may be deduced by considering each point on the moving surface of the transducer to act as a point source. Interference of the waves emitted from each point source produces a variation of acoustic pressure along the central axis of a progressive sound wave as shown in Figure 3.1.2. The final pressure maximum occurs at a distance a^2/λ from the source, where a is the radius of the source and λ is the sound wavelength. The region between the source and the final maximum is called the 'Fresnel zone', and that beyond the 'Fraunhofer zone'. In the Fresnel zone the sound beam is plane and not divergent. Although the variation of sound pressure with distance along lines parallel to the axis of the sound beam but not passing through the centre of the source differs from that shown in Figure 3.1.2, the average sound intensity at all cross-sections is constant. In the Fraunhofer zone, however, the sound beam is divergent and appears to originate at the centre of the source: the sound intensity here falls off as the square of the distance. For meaningful absorption measurements, therefore, it is necessary to work within the Fresnel zone, and at low frequencies the diameter of the transducer must be sufficiently large for this to be assured. In the Fraunhofer region not only is an

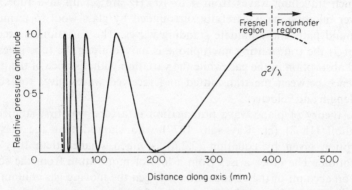

FIGURE 3.1.2. Acoustic pressure distribution along the axis of radiation for a circular transducer of radius $a = 40$ mm radiating into a medium where the acoustic wavelength λ is 4 mm.

incorrect absorption coefficient obtained because of beam spreading, but problems often arise from the reflexion of parts of the divergent beam from the boundaries of the medium in which the sound is travelling. Corrections for such effects are discussed by Bass (1958), McSkimin (1960) and Gitis and Khimunin (1969).

3.2 The Tube Method: 0·1–10 kHz atm^{-1}, Gases; 200 kHz–10 MHz, Liquids

The propagation of a plane sound wave in a closed, smooth-walled tube is used to study vibrational relaxation in gases (Chapter 4). In the standing wave system a rigid metal tube containing the gas to be investigated is excited into 'organ pipe' resonance at one of its natural frequencies by a vibrating diaphragm or ribbon loudspeaker placed at one end of the tube. A similar receiver at the other end of the tube has a response which is independent of frequency in the region of interest. The basic measurement is the band width of the bell-shaped resonance curve of the tube from which the sound absorption may be calculated. The sound velocity at a given frequency is determined by moving the loudspeaker from one stand-ing wave position to the next. Such a system has been used above 100 Hz by many workers, including Parker (1961), Henderson and Donnelly (1962) and Smith and Harlow (1963). Another arrangement is to excite such a tube at its resonant frequency and study the decay of the acoustic signal after the source of excitation is switched off. This method is similar to the reverberation method in Section 3.4, and is discussed by Edmonds and Lamb (1958) and Holmes, Smith and Tempest (1963).

An alternative system was developed by Shields and Lagemann (1957) in which travelling waves from 4 to 10 kHz are set up in a tube. The receiver microphone is carefully surrounded by glass wool to terminate the sound path and eliminate standing waves. The variation in receiver output as the transmitting microphone is moved along the tube gives the sound absorption in the gas, while the variation with distance of the phase difference between the transmitted and received waves gives the sound wavelength and velocity.

The theory of plane wave propagation in a tube was first studied by Kirchhoff (1868) (cf. Rayleigh, 1877) who showed that the classical absorption given by equation (2.2.6) is much smaller than the 'tube absorption'. The latter arises from a loss of momentum from the sound wave on account of the viscous drag between the moving gas column and the walls of the tube and also from a loss of thermal energy from the sound wave to the tube. In the case of an infinitely long, smooth, cylindrical, wide tube in which the layer of gas affected by the walls of the tube is small in comparison with the tube radius, Kirchhoff (1868) showed that for an ideal gas

$$\alpha_{\text{tube}} = \frac{(\pi \eta_s)^{\frac{1}{2}}}{r} \left[\frac{1}{\gamma^{\frac{1}{2}}} + \frac{\gamma - 1}{\gamma} \left(\frac{\kappa}{C_v \eta_s} \right)^{\frac{1}{2}} \right] \left(\frac{f}{P} \right)^{\frac{1}{2}}$$

$$= k_t \left(\frac{f}{P} \right)^{\frac{1}{2}} \tag{3.2.1}$$

while

$$U_{\text{tube}} = U - 18 \cdot 32 k_t U^2 (fP)^{-\frac{1}{2}} \tag{3.2.2}$$

Here U is the sound velocity in a tube of infinite radius and r is the tube radius.

In a tube with progressive waves, Shields and Lagemann (1957) found that the Kirchhoff equations were accurately obeyed in non-relaxing gases. Only in very low pressure gases where wall effects predominate do the Kirchhoff equations break down (Shields, Lee and Wiley, 1965; Shields and Faughn, 1969). In a tube with standing waves, however, the assumptions used by Kirchhoff (1868) are not applicable, and the observed absorption is typically 10% greater than predicted by equation (3.2.1). Thus in order to study relaxation processes in gases using the standing-wave technique, it is necessary to calibrate the tube empirically with a non-relaxing gas.

Figure 3.2.1. shows velocity dispersion and absorption curves obtained by Shields (1962) in fluorine by the travelling-wave technique. The lower frequency limit of the tube method is that at which the length of the tube equals half a sound wavelength. The high frequency limit is set by the

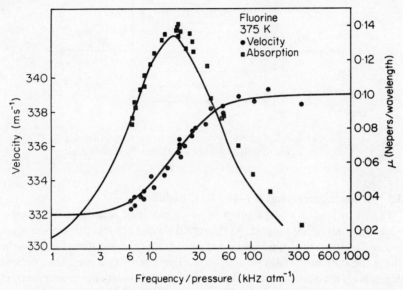

FIGURE 3.2.1. Sound absorption and dispersion in fluorine gas at 375 K determined by the tube method. The curves are calculated from equations (2.5.14) and (2.5.19) for a single relaxation process using a value of 17 kHz atm⁻¹ for the frequency to pressure ratio at which maximum absorption occurs (after Shields, 1962).

occurrence of transverse sound waves in the tube at frequencies above $0.5861 U/2r$ where r is the tube radius. Typical tube dimensions are 1·5 m length and 0·1 m internal diameter. Examples of the use of the technique are in the study of vibrational relaxation of the halogens (Shields, 1960 and 1962) and of oxygen–water mixtures (Harlow and Nolan, 1967).

A variation on the tube method has been devised by Eggers (1968) for studying liquids in the frequency range 200 kHz–10 MHz. A cylindrical resonator of the type shown in Figure 3.2.2 is driven by an X-cut quartz crystal T and the resonant frequencies of the assembly detected by R. The absorption coefficient of ultrasound in the liquid is obtained from the bandwidth of a resonance curve, while the sound velocity is calculated from the difference between two neighbouring resonant frequencies. Eggers' technique is not absolute, but requires calibration with a liquid of known properties. The main practical difficulty is in obtaining a loss-free material to bond the transducers to the liquid container. The advantage of the method is that only about 10 cm³ of liquid are required.

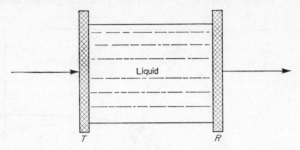

FIGURE 3.2.2. Eggers (1968) cylindrical resonator for liquids. T, exciting transducer; R, detecting transducer.

3.3   Acoustic Streaming: 0·1–10 MHz, Liquids

In this technique a plane progressive sound wave is propagated along a wide tube containing a liquid. As the sound wave is absorbed, its momentum is transferred to the liquid and a pressure gradient is produced in the tube along the direction of the sound wave. The size of this pressure gradient depends on the absorption coefficient α of the liquid. A convenient method of detecting the pressure is by direct observation of the velocity of flow of the liquid in a small side-tube connected in parallel with the main tube. Figure 3.3.1 shows the experimental arrangement used by Hall and Lamb

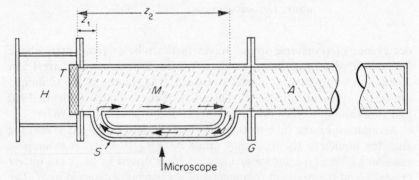

FIGURE 3.3.1. Apparatus for studying acoustic streaming in liquids: H, air-filled transducer housing; T, transducer; M, main tube containing liquid under investigation; S, side-tube; G, flanges to hold plastic diaphragm; A, tube containing absorbing liquid (after Hall and Lamb, 1959).

(1959). A transducer T of quartz or barium titanate, 60 mm in diameter with an air housing H on one side, radiates directly into a main tube M, 54 mm in diameter and 210 mm long. After passing through the liquid in M, the attenuated ultrasonic wave is absorbed in a long tube which contains a highly absorbing liquid A such as acetic acid or methyl cyclo-hexane. A thin plastic diaphragm is clamped between flanges G to separate the absorbing liquid from the test liquid. The pressure gradient in the main tube causes the liquid to flow through a glass side tube S as indicated. S has an internal diameter of 7 mm, and the flow of liquid is observed by following the motion of small suspended particles of aluminium dust with a travelling microscope.

The pressure difference ΔP between two points at distances z_1 and z_2 from the transducer is (cf. Section 2.2)

$$\Delta P = E[\exp(-2\alpha z_1) - \exp(-2\alpha z_2)] \qquad (3.3.1)$$

E, the energy density of the sound wave at the transducer, may be calculated from measurements of the electrical impedance of the transducer (Hall and Lamb, 1959). The velocity of flow v of the liquid along the side-tube of radius r and length l is given by Poiseuille's formula

$$v = \Delta P r^2 / 4\eta_s l \qquad (3.3.2)$$

These equations cannot be used directly to give α because of end effects arising from the insertion of the side-tube into the main tube. Hence equations (3.3.1) and (3.3.2) are modified to give

$$k_s \eta_s v / E = \exp(-2\alpha z_1) - \exp(-2\alpha z_2) \qquad (3.3.3)$$

where k_s is an empirical calibration constant determined from measurements of v and E in a number of liquids of known α and η_s.

This technique has not been widely used, most workers preferring the reverberation method (Section 3.4) of determining ultrasonic absorption in liquids in this frequency range. An improved method of determining the velocity of liquid flow in the side-arm is by the use of a small thermistor bead: Pigott and Strum (1967a, b) showed that when this was connected in a Wheatstone bridge circuit the unbalance voltage gave a linear measure of flow velocity. The availability of small differential pressure transducers should also improve the sensitivity of the technique: by inserting such a transducer in a side-arm, the need for a flow of liquid is removed with consequent elimination of the disturbance of the pressure profile in the tube. With flowing liquid in the side-arm, the range of absorption coefficients which may be measured is typically between $\alpha = 3 \times 10^{-2} \text{ m}^{-1}$ and 3 m^{-1}.

3.4 The Reverberation Method: 10 kHz–1 MHz, Liquids

If a sphere filled with liquid is excited into vibration, the absorption of sound in the liquid may be calculated from the rate of decay of this vibration. Figure 3.4.1 shows a typical experimental arrangement (Ohsawa and Wada, 1967). A spherical flask of thin-walled glass, some 5 litres in capacity with a narrow neck, is filled to the foot of the neck with the liquid to be investigated. The flask is supported on a wire triangle in an

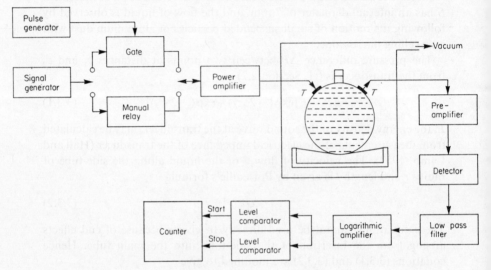

FIGURE 3.4.1. Reverberation method of studying ultrasonic absorption: T, transducer (after Ohsawa and Wada, 1967).

evacuated chamber, and two transducers T of barium titanate are cemented to the outside of the flask. As the transducers are heavily damped by this process, they may be driven at any frequency within a wide range. A pulse of radiofrequency oscillation is applied to the driving transducer: if this frequency corresponds to a resonant frequency of the liquid-filled sphere, the latter will execute a radial vibration of finite amplitude. The decay of this vibration after the driving pulse is switched off is detected by the second transducer. The resulting r.f. signal is rectified, demodulated and passed through a logarithmic amplifier. The time Δt for the decay of this signal from a value L_1 to L_2 is measured by a level comparator and counter. The rate of decay D of the excitation of the vessel is then

$$D = \frac{1}{t}\ln\frac{a}{a_0} = \frac{L_1 - L_2}{\Delta t} \tag{3.4.1}$$

where a_0 is the initial amplitude of vibration, and a that after a time t has elapsed.

A resonant radial vibration of a spherical vessel filled with liquid involves standing longitudinal waves in the liquid. The decay D of the vibration of the system is determined partly by the absorption coefficient α of these waves in the liquid, and partly by losses in the container walls and at the liquid–container interface. It can be shown that (Ohsawa and Wada, 1967)

$$D = [-(A/d)\ln(1-\beta)+\alpha]\,U \qquad (3.4.2)$$

where d is the diameter of the vessel, U is the sound velocity, A is a constant and β represents container and wall losses. β depends on the shape, wall thickness and material of the container, on the properties of the liquid in the container, and on the mode of vibration of the system. Neither A nor β can be calculated from first principles, and so this technique cannot be used to give an absolute value of α.

To make a relative determination of α by calibrating the system with a liquid of known absorption, one chooses a reference liquid with viscosity, sound velocity and density as similar as possible to those of the liquid under investigation. If β is then the same for both liquids, equation (3.4.2) yields

$$\alpha - \alpha_{\text{ref}} = \frac{D}{U} - \frac{D_{\text{ref}}}{U_{\text{ref}}} \qquad (3.4.3)$$

Such a procedure is most successful in studying dilute solutions where a comparison can be made between the solvent and the solution, as in the work of Kurtze and Tamm (1953) on dilute aqueous electrolytes. In other cases careful calibration is required with a number of different liquids: the uncertainties in this process limit the technique to systems where the major part of the decay is caused by sound absorption, and hence vessels of small diameter cannot be used with any confidence (Stuehr and Yeager, 1967). Another difficulty is in ensuring that a pure radial motion of the sphere is being excited: tangential modes are damped out quickly because of viscous losses at the interface.

Both the streaming and reverberation methods use large quantities of liquid and require calibration with liquids of known properties. At present there is no technique available which will operate satisfactorily at frequencies below 100 kHz with small liquid or solid samples.

3.5 Optical Diffraction Method: 1–100 MHz, Liquids and Gases

In 1932 Lucas and Biquard, and Debye and Sears showed that an ultrasonic wave could act as a diffraction grating for a light beam. The successive compressions and rarefactions in the sound wave periodically alter the refractive index of the medium. Although such a grating is moving with the speed of sound, the sound velocity is so much smaller than the velocity of light that the grating is effectively stationary with a grating spacing equal to the sound wavelength.

Figure 3.5.1 illustrates the experimental system. Monochromatic light is focused into a parallel beam, passes through the liquid, and is focused

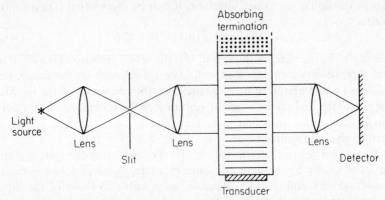

FIGURE 3.5.1. Optical diffraction method of studying ultrasonic propagation.

onto a detector such as a photomultiplier or a photographic plate. A continuous plane sound wave is passed through the liquid from the transducer to an absorbing termination. Several diffracted images of the slit are seen on either side of the central image. The absorption coefficient of the ultrasonic wave in the liquid may be determined by measuring the intensity of the diffracted light from different parts of the sound wave. The sound wavelength may be calculated from the angle of diffraction θ using the Bragg relation

$$n\lambda_1 = 2\lambda \sin \theta \tag{3.5.1}$$

where n is the order of the diffracted line, λ_1 is the light wavelength, and the grating spacing is the sound wavelength λ; the product of λ and the sound frequency gives the velocity of sound in the liquid.

Velocity determinations by this technique can easily be made to $\pm 0.1\%$, while the absorption coefficient can be measured to $\pm 3\%$ in the range 1 to $10^3\,\mathrm{m}^{-1}$. This technique is simple and convenient around room temperature. For most purposes, however, the pulse method is to be preferred for studies of liquids.

Measurements of the velocity of sound in gases have also been made by the optical diffraction method (Martinez, Strauch and Decius, 1964). This is not sufficiently sensitive to work at gas pressures much below half an atmosphere, but at higher pressures, velocities can be measured to 1 in 1000, a figure comparable to that achieved with the ultrasonic interferometer (Section 3.6). The particular advantage of the technique is that a precision mechanical system is not required (Strauch and Decius, 1966).

3.6   The Ultrasonic Interferometer: 50 kHz atm^{-1}–10 GHz atm^{-1}, Gases; 1–100 MHz, Liquids

Measurement of the velocity of sound with an ultrasonic interferometer is the most widely used acoustic technique for studying molecular energy transfer in gases (Chapter 4). The interferometer is also used to measure sound absorption in gases and sound velocity in liquids.

The single crystal ultrasonic interferometer was first devised by Pierce (1925) and is illustrated in Figure 3.6.1. It consists of a transducer T,

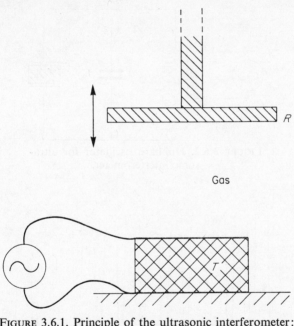

FIGURE 3.6.1. Principle of the ultrasonic interferometer: T, transducer, silvered on opposite faces; R, movable reflector.

usually an X-cut quartz crystal, which is silvered on two opposite faces and connected to an oscillator. The frequency of the oscillator is adjusted until it coincides with a natural resonant frequency of the crystal which then vibrates with appreciable amplitude. The moving surface of the crystal generates a plane sound wave which travels through the gas to a plane reflector plate R which is maintained accurately parallel to the crystal surface and can be moved normal to the crystal along the direction of the sound beam. When the distance between the transducer and the reflector is an integral number of half-wavelengths, standing waves are set up in the gas column. The reflected wave arriving back at the crystal surface is then 180° out of phase with the motion of the crystal. The resulting decrease in the amplitude of the crystal oscillations is accompanied by a decrease in the alternating current through the crystal.

Although many different detection systems have been proposed, the most satisfactory are based on a technique introduced by Hubbard (1931). The driving oscillator is loosely coupled to an LC circuit with the quartz crystal in parallel with the capacitor as in Figure 3.6.2. The oscillator and LC circuit are tuned to the resonant frequency of the crystal and the current through the crystal measured. Figure 3.6.3. shows the variation in

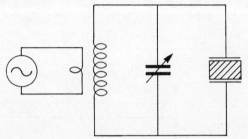

FIGURE 3.6.2. Hubbard oscillator for ultrasonic interferometer.

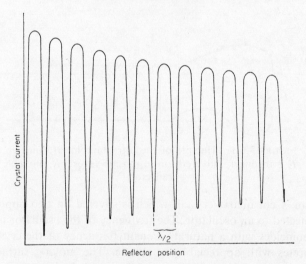

Reflector position

FIGURE 3.6.3. Variation of crystal current with reflector distance in ultrasonic interferometer.

crystal current with reflector position. The distance between successive sharp minima is half the wavelength of sound in the gas. If the frequency of the crystal oscillations is measured, the velocity of sound can then be calculated. A small correction is necessary to give the sound velocity in the ideal gas (Cottrell and McCoubrey, 1961). The absorption coefficient of the sound wave may be found from the decrease in the variation of crystal current with increasing reflector distance. Hubbard (1931) has shown that if the absorption coefficient is sufficiently small for a measurable signal to be obtained with the reflector a large number of wavelengths distant from the crystal,

$$\frac{\Delta i_A}{\Delta i_B} = \frac{\sinh 2\alpha x_A}{\sinh 2\alpha x_B} \tag{3.6.1}$$

or

$$\ln (\Delta i_A/\Delta i_B) \approx 2\alpha(x_B - x_A) \tag{3.6.2}$$

Here Δi_A is the difference between the maximum and minimum values of the crystal current for a separation x_A of crystal and reflector. For gases with high absorption coefficients, other approximations must be used (E. S. Stewart, 1946).

Two main types of mechanical system are used to provide the variable transducer–reflector separation. At frequencies below 1 MHz the crystal and reflector are contained in a large chamber having a diameter several times that of the transducer. The moving reflector shaft is introduced through a vacuum seal or bellows, and its displacement measured with a micrometer. Typical of such a system is the apparatus of Rossing and Legvold (1955). At higher frequencies the shorter acoustic wavelength requires greater mechanical precision if the reflector and crystal surface are to be maintained accurately parallel. In this case the reflector is often the end of a long piston moving in a cylinder at the other end of which is placed the transducer, as exemplified by the apparatus of J. L. Stewart (1946).

A number of precautions must be taken if meaningful results are to be obtained from the ultrasonic interferometer. The transducer and reflector must be aligned parallel to within a wavelength of light. Lack of parallelism leads to a broadening and splitting of the current minima in Figure 3.6.3: this renders velocity determinations inaccurate and absorption measurements meaningless. Diffraction effects are troublesome at low frequencies, particularly with quartz transducers where the diameter cannot readily exceed 30 mm: with solid dielectric 'Sell' transducers, the larger diameters available permit extension of the frequency range down to 100 kHz (Section 3.1). It is usual to make measurements in the Fresnel zone and to use a large diameter vessel to minimize side-wall reflections at low frequencies. Even if the reflector alignment is correct and diffraction effects minimized, however, subsidiary resonances occur in the interferometer chamber. These have been investigated theoretically by Krasnooshkin (1944) and experimentally by Bell (1950), and appear as small satellite peaks superimposed on the main resonance peaks in the crystal current. They arise from radial waves within the interferometer, and can to some extent be eliminated by packing the walls of the interferometer chamber with an acoustically absorbing medium. Stewart and Stewart (1952) have made a careful evaluation of the effect of radial waves, and conclude that it is usually necessary to apply small corrections to the observed wavelength and absorption coefficient.

Another disadvantage of an interferometer with quartz crystal transducers is that an appreciable intensity of sound is only generated at the natural resonance frequencies of the crystal. This means that the effective frequency range of the instrument is limited to the range 1–100 MHz in liquids. In gases where the number of collisions experienced by a gas molecule is proportional to the pressure, the effective frequency over pressure range of the instrument can be considerably extended by varying the pressure. Correction for non-ideality in gases is normally feasible up to at least 10 atm. Extension to low pressures is limited by the increase in sound absorption, but pressures down to 10^{-3} atm are usable especially with solid dielectric transducers (Section 3.1). The latter have the additional advantage of being non-resonant so that operation is possible over a continuously variable range of frequencies.

The wavelength of sound in a typical gas in which the sound velocity is 300 m s^{-1} is 0·3 mm at 1 MHz. A micrometer screw can be read to ± 1 μm, and so the sound wavelength can be measured to better than 1 in 1000 if measurements can be made over several wavelengths, that is if the absorption is not too great. At higher frequencies accuracy is improved by optical methods of distance measurement such as the Moire fringe technique (Blythe, Cottrell and Day, 1965), or by a differential spring system (Imai and Rudnick, 1969). Since frequency can be measured to high accuracy with an electronic counter, velocity values accurate to 1 in 1000 are easily obtained. Reproducibility is often much better than this, but the effects of transverse waves and diffraction limit the absolute accuracy, particularly at low frequencies. Absorption measurements in gases are much less accurate, errors less than 10% not normally being attainable with the single crystal interferometer. In liquids velocity measurements are limited to higher frequencies by diffraction, the accuracy being about $\pm 0·1\%$. The interferometer has not been widely used for absorption measurements in liquids, although the conditions for such studies have been reviewed by Carome, Gutowski and Schuele (1957).

Many of the difficulties with the interferometer arise from the use of standing waves. Such problems are overcome by pulse technique (Section 3.7). Alternatively, a two-transducer interferometer may be used (Greenspan, 1950; Greenspan and Thompson, 1953). In this the reflector is replaced by a second transducer which produces an electrical output proportional to the sound intensity at its surface. The interferometer is operated not in the acoustic interference region as is usual with single crystal instruments, but under conditions where the reflected intensity at the transmitting crystal is so small that standing waves are effectively absent. The amplitude of the sound wave at the receiving crystal then varies exponentially with the distance between the crystals, and the phase of the received sound wave varies linearly with path length. The sound absorption is calculated from the change in the amplitude of the received signal with distance: velocity is determined by comparing the phase of the output and input signals to the interferometer either electronically or by the simple technique of Lissajous figures. The uncertainty in the determination of the absorption coefficient in gases by this technique is less than 1%; velocities in gases can be determined up to 20 GHz atm^{-1} to an accuracy of $\pm 2\%$ by comparing the phase of a sound wave propagated through the gas with that of a sound wave of the same frequency propagated through a liquid of known ultrasonic velocity (Popov and Yakovlev, 1969).

3.7 The Pulse Technique: Liquids and Solids, 5 MHz–10 GHz; Gases, 50 kHz atm^{-1}–100 MHz atm^{-1}

This is the most widely used method of studying ultrasonic propagation in liquids and solids, while a similar technique is sometimes used in gases. It was originally developed from radar principles by Pellam and Galt (1946) and Pinkerton (1947). Figure 3.7.1 shows a typical experimental arrangement for liquids. A d.c. pulse of some 10 μs duration gates an oscillator to give a radiofrequency pulse typically of 100 V amplitude. This is applied to a transducer T which is usually a quartz crystal disc and which is bonded to a fused quartz delay line some 50 mm long. If the radiofrequency corresponds to an odd harmonic of the transducer, an ultrasonic pulse is generated. This travels through the delay line, into the liquid and then into a second delay line, at the end of which it is reconverted into an r.f. pulse by a second transducer. This pulse is amplified and demodulated. Its amplitude is measured by comparison with a reference pulse of known amplitude derived from a reference oscillator and calibrated piston attenuator: this comparison may be visual or automatic (Anderson and Chick, 1969).

The amplitude of the received pulse as a function of the distance between the two delay lines gives the absorption coefficient of sound in the liquid.

An alternative method of measuring the amplitude of the ultrasonic pulse is to put the piston attenuator in series with the acoustic system: the amplitude of the received signal is then maintained constant, an increase in attenuation from increased liquid path being compensated by decreasing

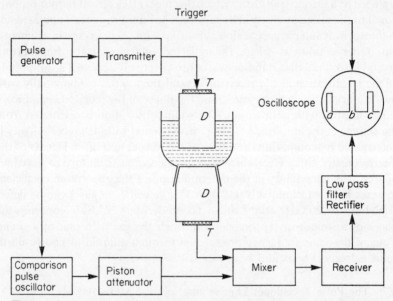

FIGURE 3.7.1. Pulse technique for studying ultrasonic propagation in liquids: T, transducer; D, delay line; a, 'breakthrough' from transmitted pulse; b, pulse which has traversed entire acoustic system; c, pulse which has made three traverses through liquid.

the attenuation from the attenuator. The sound velocity may be found by measuring electronically the time required for the ultrasonic pulse to traverse the entire acoustic system: on changing the liquid path length, the velocity is found from the difference in the transit times. Alternatively, the velocity is determined using standing waves as in the ultrasonic interferometer (Section 3.6).

If the transducers are inserted directly into the liquid to be investigated, its absorption coefficient must be sufficiently low to permit sound propagation through at least 30 mm of liquid: in this way the duration of the ultrasonic pulse is maintained less than the transit time through the liquid and standing waves are avoided. When the absorption coefficient of the liquid is high, delay lines are used to introduce an extra path length and time delay. The problem then arises of bonding the transducers to the delay line. A bond should ideally be thin and homogeneous with sufficient plasticity to allow differential thermal expansion of the transducer and delay line (Bateman, 1967; Inamura, 1970). Many materials have been proposed for this purpose including stopcock grease (McSkimin, 1957), 4-methyl pent-1-ene (Brugger and Mason, 1961), epoxy resin (Levy and Rudnick, 1962), glycerol (Wolf, 1965) and an oil (Struykov and Shchegolev, 1967).

In solids, the transducers are bonded on to opposite faces of the specimen to be investigated. The ultrasonic pulse is now reflected back and forth many times within the sample, the output signal being an exponentially decaying train of pulses. The decay in the amplitude of these pulses with increasing number of transits across the sample gives the absorption coefficient, while the transit time gives the velocity. The method is used with both longitudinal and shear waves: the detailed design of apparatus for studying solids is reviewed by Truell, Elbaum and Chick (1969).

The mechanical system for supporting the delay lines in a liquid must meet the same requirements as in the ultrasonic interferometer (Section 3.6), namely maintenance of adequate parallelism of the delay lines and accurate measurement of small displacements. Because of diffraction problems, the pulse technique is not normally used below 5 MHz in liquids or solids, while frequencies up to 300 MHz are readily accessible. Velocities may be determined to $\pm 0.2\%$ and absorption coefficients to $\pm 2\%$. Excellent reviews of the problems of constructing pulse apparatus in this frequency range are given by Andreae and coworkers (1958), Edmonds, Pearce and Andreae (1962), Andreae and Joyce (1962) and Edmonds (1966). A completely automated pulse apparatus has been described by Tabuchi (1968).

In extending the pulse technique to higher frequencies in liquids, difficulties arise from the short sound wavelength and high absorption. Although mechanical systems having the mechanical precision necessary to maintain parallelism of the delay lines and to measure small displacements have been constructed (Plass, 1965), it is more usual to use a normal incidence reflexion technique. A thin layer of liquid is placed on the end of a delay line in which sound waves are generated either by a transducer or by direct surface excitation. The change in amplitude of the sound waves on reflexion at the delay line–liquid interface gives the longitudinal impedance of the liquid: the details of this method are discussed for the shear wave case in Section 6.8. The sound velocity equals the impedance divided by the density in liquids of low absorption coefficient (cf. Section 5.3). This technique has been used up to 10 GHz by Stewart and Stewart (1963).

Absorption measurements in liquids up to 2 GHz have been made by Dunn and Breyer (1962) and Kelpin and Weis (1967). One side of a quartz crystal is surrounded by liquid, and a pulse of sound waves with a pulse length of 0·1 s generated in the liquid. The sound intensity is found by measuring the temperature rise in a miniature thermocouple probe immersed in the liquid. This can be moved along the sound path, and the change in sound intensity with distance gives the absorption coefficient. In solids, measurement of sound absorption and velocity up to 10 GHz is simpler than in liquids because of the lower absorption coefficients. The techniques are essentially similar to those at low frequencies, with signal averaging techniques sometimes used to improve the accuracy (Tittmann and Bömmel, 1967, 1968; Hemphill, 1968; McSkimin and Bateman, 1969; Williamson, 1969).

An alternative method of measuring ultrasonic velocity in liquids is the 'sing-around' method described by Greenspan and Tschiegg (1957) and illustrated in Figure 3.7.2. A pulse of sound waves from the transmitter T

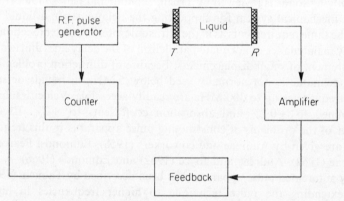

FIGURE 3.7.2. 'Sing-around' technique for measuring ultrasonic velocity in liquids. T, transmitting transducer; R, receiving transducer.

passes through the liquid to the receiving transducer R. After amplification and reshaping, the signal is fed back to the transmitting crystal. The repetition frequency of the complete cycle is determined for a given liquid path length, the period of one round trip being determined by the time delays in the electronic circuitry and in the liquid. Since the former delay is negligible, the velocity of sound in the liquid can be determined. This technique is capable of reproducibility of the order of 1 in 10^4, but imperfect pulse definition usually reduces the absolute accuracy below this value. The method is also used in solids.

Sound velocity measurements with an absolute accuracy of 1 in 3×10^4 are attainable in liquids using a phase comparison technique described by McSkimin (1957) and Barlow and Yazgan (1966). The acoustic system is shown in Figure 3.7.3. The two delay lines are maintained at a precisely fixed distance apart in the liquid by a fused quartz spacer of known thickness. A pulse of sound waves from the transducer is reflected at both the liquid interfaces as shown, and the phase difference between the waveforms in the two pulses measured. Barlow and Yazgan (1966) added each pulse

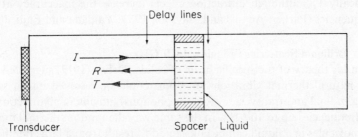

FIGURE 3.7.3. Phase comparison technique for measuring ultrasonic velocity in liquids: I, incident sound wave; R, sound wave reflected at delay-line–liquid interface; T, sound wave reflected at liquid–delay-line interface.

in turn to a continuous wave signal and adjusted the amplitude and phase of the latter to cancel exactly first one pulse and then the other. This gave the fractional part of a cycle phase difference between the two pulses, but did not indicate the number of complete cycles which forms the major part of the phase difference. The experiment was repeated at a frequency differing from the original by about 1 in 1000, and the new phase difference between the two reflected signals determined. If the number of complete cycles of phase difference in the fixed path length is the same in both cases, this may be evaluated. Hence the total phase difference or number of sound wavelengths in a given liquid path is known accurately, and the product of wavelength and frequency gives the velocity. Sound absorption can also be determined with the arrangement in Figure 3.7.3 by comparing the amplitude of the pulses reflected from the two interfaces. The phase comparison technique has been applied over the frequency range 10–200 MHz. A variation of the method that gives accurate velocity measurements in solutions by comparison with the solvent has been described by Schaaffs and Kalweit (1960), Kaulgud (1965) and Bader (1968).

The so-called 'direct' method of measuring sound absorption in gases is a pulse technique similar to that used in liquids. Because of the lower sound velocity, a transducer can be inserted directly into the gas and delay lines are not required to give adequate pulse lengths. Two transducers, either quartz crystals or 'Sell' solid dielectric transducers, are employed, and the sound intensity at the receiving transducer is measured as a function of sound path length to give the absorption coefficient. It is usual to operate at frequencies between 50 and 500 kHz, although the effective frequency range may be greatly extended by altering the gas pressure. The elimination of standing waves considerably improves the accuracy of this technique relative to the ultrasonic interferometer, the error in α being typically 1%, although diffraction effects increase the inaccuracy at low frequencies (Parker, Adams and Stavseth, 1953; Yamada and Fujii, 1966).

3.8   Brillouin Scattering: Liquids, 1–10 GHz

In his theory of the specific heats of solids, Debye (1912) suggested that the natural thermal vibrations of atoms could be resolved into a set of longitudinal and transverse waves (or phonons) moving in all directions at all frequencies up to 10^{13} Hz. In the same way the random thermal motion of molecules in a liquid can be considered to result from a large number of longitudinal waves.

At low sound frequencies it is convenient to use a transducer to superimpose a sound wave of definite frequency on the random sound waves already present in the liquid. At hypersonic frequencies ($> 10^9$ Hz), however, the efficiency of transducers is so low that it is preferable to study the sound waves naturally present in the liquid. Brillouin (1922) suggested that this could be done by investigating the scattering of light by the liquid. As a light wave passes through the liquid, it generates an oscillating electric dipole at each point. This radiates electromagnetic energy in every direction, but if the liquid is perfectly homogeneous the vector sum of the energy scattered in any direction except the forward is zero. The thermal motion of the molecules leads to inhomogeneities and to a periodic modulation of the scattering power of the liquid. Hence the light beam suffers a Bragg reflexion from the acoustic diffraction grating

$$n\lambda_1 = 2\lambda \sin \theta \qquad (3.5.1)$$

Here λ_1 is the light wavelength, λ the sound wavelength and θ the angle between the incident light and the plane of the grating.

Since the light is reflected by a moving sound wave, it suffers a Doppler shift. The magnitude of this frequency shift is determined by the component of the sound velocity in the direction of the light beam $U\cos\left[(\pi/2) - \theta\right]$. Thus the Doppler shift $\Delta\nu$ of light of frequency ν_1 is

$$\frac{\Delta\nu}{\nu_1} = \pm \frac{2U}{U_1} \sin \theta \qquad (3.8.1)$$

where U is the velocity of sound and U_1 the velocity of light. The scattered light consists of a doublet split symmetrically about the incident frequency. Combination of equations (3.5.1) and (3.8.1) shows that the separation of the scattered light from the incident light equals the frequency of the sound wave responsible for the scattering. The wavelength of the sound wave is given by equation (3.5.1) and so the velocity of the sound wave can be calculated. By considering the detailed theory of scattering, a relation may be obtained for the intensity of the scattered light and for the shape and width of the Brillouin lines. From the latter, the absorption coefficient of the sound wave in the liquid may be found.

The intensity of the scattered light is very small, and the displacement of the Brillouin component from the incident beam is less than 10 GHz in a liquid. Hence an intense monochromatic light source is required with an optical interferometer for detection. The technique has proved extremely difficult with conventional light sources (Venketeswaran, 1942; Fabelinskii, 1957). With the advent of the laser, however, these difficulties were overcome, and reliable values of sound velocity can be obtained with an accuracy better than $\pm 0.1\%$.

Figure 3.8.1 shows a typical experimental arrangement for studying spontaneous Brillouin scattering in liquids (Fleury and Chiao, 1966). The

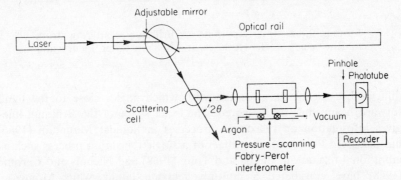

FIGURE 3.8.1. Apparatus for observing Brillouin scattering as a function of angle. The scattering angle θ is varied by a combination of rotation of the adjustable mirror and translation along the optical rail (after Fleury and Chiao, 1966).

scattering angle, and hence the frequency of sound observed, can be varied by moving the adjustable mirror along an optical bench. The necessary optical resolution is obtained with a pressure scanned Fabry–Perot interferometer. The main requirement for accuracy in the velocity determination is that the scattering angle θ be known precisely. Careful temperature control and dust-free liquids are also important.

Figure 3.8.2 is an interferogram of the Brillouin scattering from toluene (Fleury and Chiao, 1966). The scattered lines are displaced from the central Rayleigh peak by 3586 MHz, and this is the frequency of the observed sound wave. The scattering angle is $(70·0 \pm 0·1)°$, and the wavelength and velocity of the sound wave may be calculated from equation (3.5.1). The sound absorption is derived from the width of the Brillouin peaks: the accuracy in the absorption values is only about $\pm 10\%$ however, as it is difficult to separate the true Brillouin broadening from the instrumental broadening (Leidecker and LaMacchia, 1968).

The shortest sound wavelength available corresponds to half the wavelength of the incident light which is equivalent to a sound wave of about 10^{10} Hz. Accurate sound velocity measurements below 10^9 Hz are

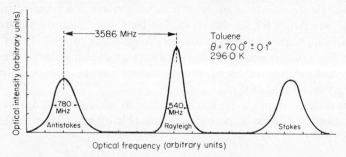

FIGURE 3.8.2. Interferogram of Brillouin scattering from toluene, using a He–Ne laser at 632·8 nm (after Fleury and Chiao, 1966).

difficult because of the precision with which angles close to the main forward beam must be measured. The exact shape of the Brillouin lines gives information on relaxation processes in liquids: Mountain (1966) has discussed the broadening effect of a single relaxation process such as vibrational relaxation, Bhatia and Tong (1968) and Nichols and Carome (1968) have considered a multiply relaxing liquid, while Montrose, Solovyev and Litovitz (1969) have discussed the calculation of the distribution of structural relaxation times from Brillouin line shapes. It is likely that the Brillouin scattering technique of investigating sound propagation in liquids will find increasing application as a means of extending the frequency range beyond that available with conventional methods of generating ultrasonic waves.

A variation in this method of measuring hypersonic velocity is by the technique of stimulated Brillouin scattering, first detected by Chiao, Townes and Stoicheff (1964). Instead of using a continuous wave laser as in spontaneous Brillouin scattering, a high power, pulsed laser is used. Within the liquid there will occur sound waves with a wavelength such that spontaneous Brillouin scattering will occur in the backwards direction along the path of the incident light. This superposition of backscattered and incident light induces by electrostriction a pressure (sound) wave of appreciable intensity. The frequency of this sound wave equals the difference in frequency between the incident and backscattered Brillouin light and hence equals the frequency of the sound waves originally responsible for the scattering. As the process continues, the intensity of the sound waves and of the scattered light increases until the scattered intensity becomes almost equal to the incident intensity. The incident and scattered light beams are sampled by an optical system of partly silvered mirrors, and the frequency difference between the beams determined in an optical interferometer. With suitable precautions the values of sound velocity obtained by stimulated Brillouin scattering agree with those from spontaneous Brillouin scattering. An advantage of the former technique is the high intensity of the scattered light. Moreover this intensity is produced during the period of the laser pulse (ca. 10^{-9} s) and no electronic or temperature stabilization is necessary (Goldblatt and Litovitz, 1967). Stimulated Brillouin scattering has also been studied in gases (Madigosky, Monkewicz and Litovitz, 1967).

3.9 The Optic–Acoustic Effect: Gases, 1 kHz atm^{-1}–1 MHz atm^{-1}

In 1881 Tyndall discovered that when periodically interrupted infrared radiation was shone on certain gases, a sound wave was emitted of frequency equal to the frequency of interruption of the radiation. Gorelik (1946) suggested that this effect could be used to measure the rate of energy transfer from vibration to translation. Energy is absorbed by an infrared active molecular vibration, and after the vibrational relaxation time has elapsed this excess vibrational energy is transferred to translation in a collision with another molecule (Chapter 4). This periodic change of temperature leads to a periodic change of pressure in a constant volume system, that is, to a sound wave.

The time which elapses between the absorption of a vibrational quantum and its transfer to translation will determine the phase difference ϕ between the modulation of the incident infrared radiation and the resultant sound wave. The vibrational–translational relaxation time at 1 atm τ may be calculated either from this phase difference or from the amplitude ξ of the sound wave by the approximate expressions (Read, 1968)

$$\tan\left[(\pi/2) - \phi\right] = \omega\tau \tag{3.9.1}$$

and

$$\xi = A/\omega C_v (1 + \omega^2 \tau^2)^{\frac{1}{2}} \tag{3.9.2}$$

where A is a constant and C_v is the total heat capacity of the gas. Figure 3.9.1 shows the variation of the phase shift and sound amplitude as a function of $\omega\tau$: at low values of $\omega\tau$ the amplitude decreases linearly with increasing frequency while at high values of $\omega\tau$ it decreases as the square of frequency. It is often convenient experimentally to vary τ as well as ω: this is achieved by changing the gas pressure P since the vibrational relaxation time is inversely proportional to the number of molecular collisions and hence to P.

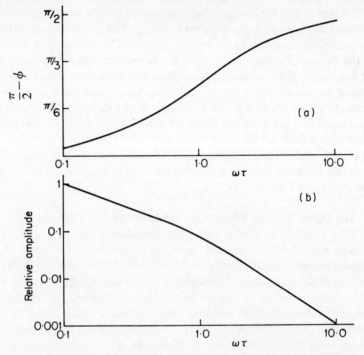

FIGURE 3.9.1. Variation (a) of phase shift ϕ and (b) of relative signal amplitude in the spectrophone as a function of frequency.

Figure 3.9.2 shows a typical arrangement of a spectrophone which utilizes the optic–acoustic effect to study vibrational relaxation in gases. A beam of infrared radiation is modulated by a chopping disc before passing into the spectrophone cell. The resulting acoustic signal in the gas in the cell is detected by a microphone: the signal amplitude is measured directly while the phase is obtained by comparison with a reference signal from a photocell.

A typical condenser microphone has a flat amplitude–frequency response up to about 10 kHz, although microphone phase shifts are introduced at frequencies above about 2 kHz, i.e. $\omega \sim 10^4$ Hz. At pressures below about 10^{-2} atm in a typical spectrophone cell an appreciable fraction of the excess vibrational energy is lost in wall collisions and equations (3.9.1) and (3.9.2) are no longer valid. Hence the lower limit of the vibrational relaxation time which can be measured is about 10^{-6} s. The upper limit occurs when an appreciable fraction of the vibrationally excited molecules are deactivated by radiation rather than collision, say 10^{-3} s (Read, 1968).

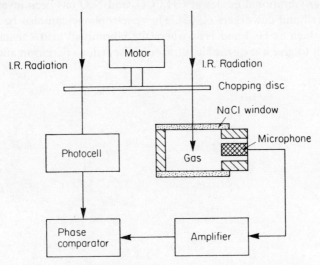

FIGURE 3.9.2. Typical arrangement of the spectrophone.

Somewhat longer times can be attained by a fluorescence competition method. If a gas has a long vibrational relaxation time, almost all of the excess vibrational energy will be lost by radiation. Only a small optic-acoustic signal will be obtained, the amplitude of which is proportional to the energy lost by collision:

$$\text{initial signal} \propto \quad p_{10}\, n_1\, h\nu \; \text{s}^{-1}\, \text{m}^{-3} \tag{3.9.3}$$

p_{10} is the probability of vibrational–translational energy transfer in a collision, n_1 is the number of vibrationally excited molecules in unit volume, and ν is the vibrational frequency. If a small amount of an impurity which is very efficient at promoting vibrational–translational energy transfer is added, the amplitude of the signal will increase because all of the excess vibrational energy is now degraded by collision. Hence

$$\text{final signal} \propto \quad (p_{10}+p_{\text{rad}})\, n_1\, h\nu \; \text{s}^{-1}\, \text{m}^{-3} \tag{3.9.4}$$

p_{rad}, the probability of a radiative transition, can be calculated from spectroscopic data. p_{10} is thus obtained from the ratio of initial to final signal amplitudes in the spectrophone, and the vibrational relaxation time may be calculated using equations (4.5.4) and (4.5.5).

Since ultrasonic techniques for studying vibrational relaxation in gases are usually limited to the lowest vibrational mode, one of the chief advantages of the spectrophone is that it can be used to study energy transfer from any desired infrared active vibration. For example, the relaxation of the higher vibrational modes of CH_4, CO_2 and N_2O has been investigated by Cottrell and coworkers (1966). The spectrophone can also be applied to gases such as HCl and HBr where the vibrational heat capacities are too small to give a measurable ultrasonic dispersion (Ferguson and Read, 1967).

CHAPTER 4

MOLECULAR ENERGY TRANSFER IN GASES

4.1 Introduction

One of the earliest applications of molecular acoustics was to the study of molecular energy transfer in gases. Energy in gases is distributed among the translational, vibrational, rotational and electronic degrees of freedom. In most molecules the electronic quanta are so large that only the ground electronic state is populated at temperatures below 10,000 K. Hence the thermodynamic properties of gases can be discussed by considering the contributions of translation, vibration and rotation only. These modes are in equilibrium with each other, and in this chapter we shall be concerned with the molecular mechanism for the transfer of energy among them. Such studies are of fundamental importance in the understanding of reaction kinetics.

The radiative lifetime of a vibrational state is of the order of 10^{-2} s, while that of a rotational state is even longer (Cottrell, 1965). Since the time between collisions for a given molecule in the gas phase at s.t.p. is typically 10^{-9} s, energy transfer will normally occur exclusively by collision and not by radiation. This is supported by experimental observations that the vibrational relaxation time is inversely proportional to the number of collisions a molecule experiences, that is to the gas pressure.

Intermolecular exchange of translational energy occurs at every collision. The exchange of energy between translation and the vibrational and rotational modes occurs less readily. For a molecular collision to be effective in changing the vibrational energy of a molecule, the duration of the collision should be comparable to the period of the vibration. A typical molecular relative translational velocity at room temperature is 5×10^2 m s^{-1}, while the effective range of the intermolecular forces during a collision is of the order of 10^{-10} m. Hence the time which two molecules spend closer than 10^{-10} m, that is the duration of the collision, will be 4×10^{-13} s. A typical molecular vibration frequency of 3×10^{13} Hz has a period of about 3×10^{-14} s. Thus only those molecular collisions in which the relative translational velocity is much higher than average will lead to vibrational–translational energy transfer. The probability of the transfer of a vibrational quantum in a collision will increase with increasing temperature and translational velocity, with decreasing vibration frequency, and with increase in the steepness of the intermolecular potential energy function.

3

The period of a typical molecular rotation is much closer to the duration of a collision, and rotational–translational energy transfer in a collision has a higher probability. The ease of rotational energy transfer can also promote vibrational energy transfer through the sequence

$$\text{translation} \rightleftharpoons \text{rotation} \rightleftharpoons \text{vibration}.$$

The two acoustic techniques which are most widely used to study molecular energy transfer in gases are the ultrasonic interferometer (Section 3.6) and the spectrophone (Section 3.9). In the former, the velocity dispersion or the change in α/f^2 with frequency give the relaxation time for intermolecular energy transfer as discussed in Section 2.5. Ultrasonic techniques are particularly useful for studying vibrational relaxation, although systems using quartz transducers are normally limited to temperatures below 846 K, the Curie temperature of quartz: at higher temperatures the shock tube method is preferred (Bradley, 1962). Ultrasonic studies of rotational relaxation are being superseded by thermal conductivity measurements (O'Neal and Brokaw, 1963) from which the probability of rotationally inelastic collisions may be deduced using the theory of Mason and Monchick (1962). The only gas for which the contribution of the electronic degrees of freedom to the heat capacity is sufficient for ultrasonic dispersion to be observed is NO (Bauer and Sahm, 1965): 70 collisions are required for electronic–translational energy transfer at 296 K.

In the following three sections the molecular theories of vibrational–translational, rotational–translational and vibrational–rotational energy transfer are outlined. Section 4.5 contains a representative selection of the available experimental results. The remaining sections provide a comparison of the theoretical predictions with the experimental results and a discussion of vibrational–vibrational energy transfer.

4.2   Theories of Vibrational–Translational Energy Transfer

Molecular theories of vibrational–translational energy transfer have been developed by considering the collision process classically or quantum mechanically. In the classical treatment it is assumed that both the translational and vibrational motions of the colliding molecules can be described by classical mechanics. In the semi-classical calculation the vibrations are treated quantum mechanically while the translational motion is described classically. The simplest physical model is an end-on collision of an atom with a diatomic molecule, and it is required to calculate the probability of the transfer of energy between translation and vibration as a result of this collision. The semi-classical treatment to be discussed is based on Cottrell and McCoubrey (1961): it is also applicable to energy transfer in a collision of two diatomic molecules since experimental results show that the probability of energy transfer to the vibrational mode of a diatomic molecule is not significantly affected by the internal structure of the collision partner.

Time-dependent perturbation theory shows that the probability of a transition under the influence of a given perturbation is proportional to the square of the matrix element of the perturbation between the states (Pauling and Wilson, 1935). If $V(t)$ is the perturbing potential which depends on time, the probability of a transition from state n to state m is

$$p_{nm} = \frac{1}{\hbar^2} \left| \int_{-\infty}^{\infty} V_{mn} \exp(2\pi i \nu_{mn} t) \, dt \right|^2 \qquad (4.2.1)$$

Here $h\nu_{mn}$ is the amount by which the internal energy is changed, and V_{mn} is the matrix element of the potential. We assume that the perturbation may be written

$$V_{mn} = x_{mn} F(t) \qquad (4.2.2)$$

where x is the displacement of the oscillator and F is the force acting on it. If zero time is taken at the distance of closest approach of the colliding particles, $F(t)$ is an even function, and the probability of de-excitation of the first vibrational level in a collision is

$$p_{10} = 4x_{10}^2 I^2 / \hbar^2 \qquad (4.2.3)$$

where

$$I = \int_0^{\infty} F(t) \cos 2\pi \nu t \, dt \qquad (4.2.4)$$

Thus in order to calculate I, we require the force–time relationship for the two colliding molecules.

The equation of motion of the colliding species is

$$F(r) = m\ddot{r} \qquad (4.2.5)$$

where m is the reduced mass of the system and r is the distance between the atom and the molecule. The combination of this force–distance relation with the potential energy function

$$V(r) = \int_r^{\infty} F(r) \, dr \qquad (4.2.6)$$

gives the time–distance relation

$$t(r) = \int_{r_0}^{r} [v^2 - (2/m) V(r)]^{-\frac{1}{2}} \, dr \qquad (4.2.7)$$

where r_0 is the distance of closest approach of the two species and v is their relative velocity at infinite separation. It is assumed that the dynamics of the collision are unaffected by energy transfer and that the translational energy of the colliding molecules is much greater than the vibrational quantum transferred: this condition is valid for the high energy collisions which are important in energy transfer.

From the expressions for $F(r)$ and $t(r)$ in equations (4.2.5) and (4.2.7), the force–time relationship $F(t)$ and the integral I may be obtained. The exact value of I depends on the form which is assumed for the potential energy function $V(r)$. If for mathematical convenience we assume that both the attractive and repulsive intermolecular forces are exponential,

$$V(r) = \lambda \exp(-\alpha r) - \mu \exp(-\alpha r/2) \tag{4.2.8}$$

we obtain

$$I = \frac{4\pi^2 m v}{\alpha} \exp\left[-\frac{2\pi v(\pi - 2\phi)}{\alpha v}\right] \tag{4.2.9}$$

where ϕ is $\tan^{-1}(\mu^2/2mv^2 \lambda)^{\frac{1}{2}}$. If the vibrational states are assumed to be those of a harmonic oscillator, the square of the displacement of the oscillator is

$$x_{10}^2 = \hbar/4\pi M v \tag{4.2.10}$$

where M is the reduced mass of the oscillator. Hence

$$p_{10} = \frac{32\pi^4 m^2 v}{hM\alpha^2} \exp\left[-\frac{4\pi v(\pi - 2\phi)}{\alpha v}\right] \tag{4.2.11}$$

or, if the attractive component of the intermolecular force is neglected by putting $\mu = 0$,

$$p_{10} = \frac{32\pi^4 m^2 v}{hM\alpha^2} \exp\left[-\frac{4\pi^2 v}{\alpha v}\right] \tag{4.2.12}$$

Equation (4.2.12) was first derived by Landau and Teller in 1936. This expression for the probability p_{10} of vibrational de-excitation of the first vibrational level as a function of the relative translational velocity v of the colliding molecules must now be averaged over the Maxwell distribution of molecular velocities in a gas. For one-dimensional motion along the line of the oscillator this is

$$dN = Nv\,dv\frac{m}{kT}\exp\left(-\frac{mv^2}{2kT}\right) \tag{4.2.13}$$

where N is the total number of molecules. Hence

$$\begin{aligned}
p_{10} &= \frac{32\pi^4 m^2 v}{hM\alpha^2 N} \int_0^\infty \exp\left(-\frac{4\pi^2 v}{\alpha v}\right) dN \\
&= \frac{32\pi^4 m^3 v}{hM\alpha^2 kT} \int_0^\infty v\exp\left(-\frac{4\pi^2 v}{\alpha v} - \frac{mv^2}{2kT}\right) dv
\end{aligned} \tag{4.2.14}$$

This integral cannot be evaluated analytically. An approximation to p_{10} can be obtained by observing that whereas p_{10} increases with increasing relative velocity v, the number of molecules with a given value of v decreases as v increases at large v. Hence p_{10} will have a maximum at a certain relative velocity v^*. Landau and Teller (1936) assumed that only molecules with velocity close to v^* will be effective in promoting energy transfer. v^* will occur at the minimum value of the negative exponent in equation (4.2.14). This occurs where

$$-\frac{4\pi^2 v}{\alpha v^{*2}} + \frac{mv^*}{kT} = 0 \qquad (4.2.15)$$

so that

$$v^* = \left(\frac{4\pi^2 vkT}{\alpha m}\right)^{\frac{1}{3}} \qquad (4.2.16)$$

p_{10} averaged over all collisions can now be calculated from equation (4.2.14) using the value of the velocity at which a transition is most likely v^* from equation (4.2.16):

$$p_{10} = \frac{64\pi^5 m^2 v}{hM\alpha^2}\left(\frac{mv^2}{9\alpha^2 kT}\right)^{\frac{1}{3}} \exp\left[-\left(\frac{54\pi^4 v^2 m}{\alpha^2 kT}\right)^{\frac{1}{3}}\right] \qquad (4.2.17)$$

This expression gives the overall probability of vibrational de-excitation of a diatomic molecule in an end-on collision with an atom for an exponential repulsive potential. It shows that if the exponential term is dominant the logarithm of the vibrational relaxation time is proportional to the two-thirds power of the vibration frequency, the one-third power of the reduced mass of the colliding species, the inverse two-thirds power of the coefficient of the repulsive potential, and the inverse one-third power of the temperature.

The probability of energy transfer in such a collision may also be calculated by a fully quantum mechanical treatment. This was originally done for the one-dimensional case by Zener (1931) and Jackson and Mott (1932), and subsequently by Schwartz, Slawsky and Herzfeld (1952). The sequence of events in an isolated binary collision is no longer considered, and the approaching and retreating atoms are instead represented by waves of wavelength λ_B given by the de Broglie equation

$$\lambda_B = h/mv \qquad (4.2.18)$$

The main component of the scattered wave will have the same wavelength as the incident wave as a result of elastic collisions, whereas the components of the scattered wave arising from inelastic collisions will have different wavelengths. The probability of energy transfer to or from a particular vibrational level of the molecule is given by the square of the amplitude of the scattered wave with the appropriate wavelength.

The details of the calculation are given by Herzfeld and Litovitz (1959) and by Cottrell and McCoubrey (1961). For the end-on collision of an atom with a diatomic molecule under the influence of an exponential repulsive potential, the quantum mechanical treatment gives

$$p_{10} = \frac{32\pi^4 m^2 \nu}{hM\alpha^2} \frac{\sinh q_i \sinh q_n}{(\cosh q_i - \cosh q_n)^2} \qquad (4.2.19)$$

where $q_i = 4\pi^2 mv_i/h\alpha$, $q_n = 4\pi^2 mv_n/h\alpha$ and v_i and v_n are the relative translational velocities before and after collision and energy transfer. Equation (4.2.19) is very similar to equation (4.2.12) derived by the semi-classical treatment. Cottrell and McCoubrey (1961) have shown that

$$\frac{p_{10}(QM)}{p_{10}(SC)} = \frac{\sinh q_i \sinh q_n}{\sinh^2 \frac{1}{2}(q_i + q_n)} \qquad (4.2.20)$$

For velocities above the thermal average, $q_i \approx q_n$ and the semi-classical and quantum mechanical transition probabilities are equal. For low velocity collisions, the semi-classical treatment is in error in assuming that the relative translational velocity is unchanged in an inelastic collision: this error is insignificant, however, since low velocity collisions are unimportant in molecular energy transfer. The quantum mechanical treatment can also be extended to include the effect of attractive forces.

These treatments of one-dimensional head-on collisions have been extended to three-dimensional collisions in which the two interacting molecules may have any relative orientation. The principles of the calculation are the same as in the one-dimensional approach, but the mathematical complexities are such that analytical solutions cannot always be obtained. The semi-classical approach was first extended to three-dimensional collisions of polyatomic molecules by Cottrell and Ream (1955) while a three-dimensional quantum mechanical treatment for linear molecules was given by Schwartz and Herzfeld in 1954: the latter was extended to polyatomic molecules by Tanczos (1956), and Takayanagi (1952, 1965) developed a similar treatment. The details of the calculations are discussed by Herzfeld and Litovitz (1959) and by Cottrell and McCoubrey (1961). The results for the transition probability with an exponential repulsive potential are similar to those for the one-dimensional case: in particular the three-dimensional semi-classical treatment of Cottrell and Ream (1955) has the same exponent and a similar pre-exponential factor to the one-dimensional expression of equation (4.2.12). This similarity arises because the probability of energy transfer is much larger for head-on collisions than for collisions with unfavourable trajectories (Shin, 1968).

The basic theory outlined above breaks down for high energy collisions in which p_{10} is predicted to be greater than unity. This absurd situation arises from the assumption inherent in perturbation theory that the transition probability in a collision is small. Although p_{10} averaged over all collisions is small, the probability of energy transfer in a high velocity collision is large and it is just these collisions which are mainly responsible for vibrational energy transfer. Rapp and Sharp (1963) showed that in molecules with high vibrational frequencies such as nitrogen this effect is only important at temperatures above 5000 K. But in molecules with lower vibrational frequencies or with strong attractive forces, significant theoretical errors can arise. In many calculations this problem is arbitrarily eliminated by assuming a constant value of $p_{10} = 0.5$ for collisions in which the theory predicts a larger value of p_{10}.

Theories of vibrational energy transfer are limited chiefly by the difficulty of solving the relevant equations and by a lack of knowledge of the exact details of the intermolecular potential. The former difficulty can be overcome by exact numerical computation of the transition probability without attempting to obtain an analytical solution. This has been done by Marriott (1964), Alterman and Wilson (1965) and Kelley and Wolfsberg (1966a, b) although at present the length of the computation allows only the one-dimensional case to be evaluated. The problem of the potential energy function has been studied by Mies (1965) who used an exact potential energy surface for the $He + H_2$ system to calculate the transition probability. A comparison of Mies' results with the conventional semi-classical treatment showed that p_{10} is critically dependent on the details of the intermolecular potential and that an exponential potential acting between atomic centres has considerable limitations when combined with the other assumption that the amplitude of vibration is small compared to the range over which the repulsive potential is effective. Although many investigations of different intermolecular functions have been made, Mies' (1965) work shows the inadequacy for energy transfer calculations of functions originally developed to describe the low energy collisions which are important in processes such as gas viscosity.

4.3 Theories of Rotational–Translational Energy Transfer

In most molecules the rotational and translational velocities are comparable and rotational–translational energy transfer in a collision may be expected to occur readily. The size of the rotational quantum ranges from 10^{-21} J in H_2 to 7×10^{-25} J in I_2. With the exception of H_2, molecules occupy many rotational levels at room temperature where

$kT \approx 4 \times 10^{-21}$ J: this means that, unlike vibrational relaxation, many different relaxation processes can occur in rotational relaxation and the observed rotational relaxation time will be some average of the relaxation times for each of these processes.

The high rotational velocity in H_2 makes rotational–translational energy transfer relatively inefficient, and perturbation methods such as were used in treating vibrational relaxation (Section 4.2) can be used to calculate the transition probability in a collision (Beckerle, 1953; Brout, 1954; Takayanagi, 1957, 1965; Davison, 1962; Roberts, 1963). In other gases, however, the transition probability in a collision is sufficiently high for a perturbation treatment to be inapplicable and calculations of the rate of energy transfer are usually based on classical considerations.

An example of a classical calculation of the probability of rotational–translational energy transfer in a collision is that of Widom (1960). He considered a monatomic gas in which some spherical-top molecules were dilutely dispersed. The atom–molecule collisions were assumed to be impulsive and the surfaces of the molecules were assumed to be rough: the criterion of roughness was that the vector representing the velocity difference between the atom and its point of impact on the surface of the molecule changed sign on collision. From a three-dimensional analysis of the trajectories and velocities of the colliding species before and after collision, Widom calculated the average change in rotational energy in a collision, and showed that the average number of collisions required for rotational energy transfer was

$$Z_r = 3(1 + b^2)/8b \tag{4.3.1}$$

where $b = I/\mu a^2$, I is the moment of inertia of the molecule, μ is the reduced mass of the colliding species and a is the sum of the atomic and molecular radii. For CH_4–Ne collisions, for example, equation (4.3.1) predicts that $Z_r = 15$ and is independent of temperature. A similar expression was derived by Wang Chang and Uhlenbeck (1951, 1964) for rotational energy transfer between identical spherical molecules:

$$Z_r = 3(1 + 2b)/8b\zeta \tag{4.3.2}$$

where ζ is a measure of the roughness of the molecular surface; $\zeta = 1$ for rough spheres as defined above and $\zeta = 0$ for perfectly smooth spheres where there is no frictional torque on collision. If $\zeta = 1$, equation (4.3.2) gives a value of $Z_r = 18$ for pure CH_4.

Sather and Dahler (1961, 1962) extended these calculations to slightly prolate spheroids and to loaded spheres, and also considered the effect of attractive intermolecular forces. Parker (1959) used a more realistic intermolecular potential but restricted collisions to a plane: he found a small increase in Z_r with increasing temperature, whereas a similar treatment by Nyeland (1967) predicts that Z_r should be independent of temperature. The effect of permanent dipoles on rotational energy transfer was considered by Zeleznik (1967) who found that Z_r decreased as the molecular dipole moment increased.

The most detailed treatment of the temperature dependence of the rotational relaxation time is that of Raff and Winter (1968) who explicitly consider the multilevel nature of molecular rotation. The spacing of the rotational energy levels of a rigid rotor increases as the rotational quantum number J increases and this leads to a longer relaxation time for $\Delta J = \pm 1$ rotational transitions at high J values. As the temperature increases, high J states become more heavily populated and the observed 'rotational relaxation time', which is an average over all J states, should increase. At very low temperatures where only two rotational levels are occupied, however, the normal decrease of rotational relaxation time with increased temperature is expected. Similar conclusions have been reached by Brau and Jonkman (1970).

4.4 Theories of Vibrational–Rotational Energy Transfer

In 1962 Cottrell and Matheson suggested that in molecules with low moments of inertia, the transfer of energy from vibration to rotation and thence to translation could occur more readily than direct vibrational–translational energy transfer. They argued that vibrational–rotational energy transfer would be preferred when the peripheral velocity of rotation of a molecule is greater than its translational velocity, and that this mechanism should be especially favoured when the vibration involves bond bending as in Figure 4.4.1.

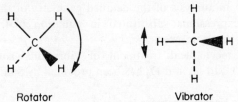

Rotator   Vibrator

FIGURE 4.4.1. Vibrational–rotational energy
transfer in gaseous methane.

Under such circumstances the probability of de-excitation of the first vibrationally excited level will be

$$p_{10} = Z \int_0^\infty p_{10}(v_r) f(v_r) \, dv_r \qquad (4.4.1)$$

where Z is the collision frequency of the gas molecules, $p_{10}(v_r)$ is the probability of vibrational–rotational energy transfer for a given peripheral velocity of rotation v_r and $f(v_r)$ is the distribution function of rotational velocities. In order to calculate $p_{10}(v_r)$, the semi-classical expression for translational–vibrational energy transfer given in equation (4.2.12) was used, with the relative translational velocity of the colliding molecules replaced by the peripheral velocity of rotation v_r: because of the large change in rotational velocity, it is necessary to symmetrize the rotational velocity by using a mean value before and after transfer of a vibrational quantum. The distribution function of rotational velocities may be obtained from the standard treatment of rotational energy levels (Herzberg, 1945). Substitution of these values of $p_{10}(v_r)$ and $f(v_r)$ into equation (4.4.1) gives an expression for the probability of vibrational–rotational energy transfer in a collision which can be integrated numerically. A similar treatment was given by Moore (1965).

A more rigorous approach to vibrational–rotational energy transfer is that of Cottrell and coworkers (1964). They considered that the relative molecular translational velocities were the same as in an ordinary collision while the rotating molecules were assumed to have the classical angular rotational velocity given by

$$\tfrac{1}{2}I\omega^2 = kT \tag{4.4.2}$$

Although the translational velocity determines the overall duration of the collision, the average rotational velocity during the collision is assumed to determine the probability of energy transfer. For the forces between the peripheral atoms of the two molecules, both a repulsive r^{-12} Lennard–Jones function and an exponential repulsion were considered. The time dependence of the intermolecular force field resulting from rotation was evaluated from an analysis of the detailed geometry of the collision, and the resulting expressions substituted in equation (4.4.1) to give the probability of energy transfer.

A quantum mechanical treatment of vibrational–rotational energy transfer from CO_2 to H_2 and D_2 has been provided by Sharma (1969).

4.5 Results for Energy Transfer in Gases

A large number of studies of vibrational relaxation in gases have been carried out by acoustic techniques. Figure 4.5.1 shows the dispersion of ultrasonic velocity arising from vibrational relaxation in SF_6 at 284·2 K (Haebel, 1968). The line through the experimental points is the single relaxation equation (2.5.14) corresponding to the relaxation of the whole of the vibrational heat capacity. The ultrasonic velocity U_0 at sound frequencies below the relaxation region is given by equation (2.1.9) with the whole of the vibrational heat capacity, while the velocity U_∞ above the relaxation region includes the contribution from translation and rotation only. Figure 4.5.2 shows the ultrasonic absorption in SF_6 at the same temperature (Haebel, 1968): the curve again corresponds to the single

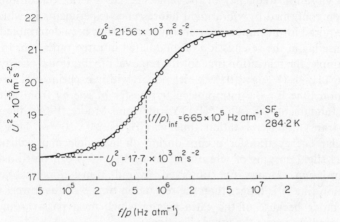

FIGURE 4.5.1. Dispersion of ultrasonic velocity in SF_6 at 284·2 K (after Haebel, 1968).

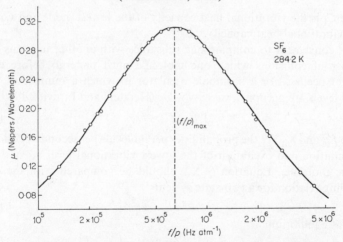

FIGURE 4.5.2. Ultrasonic absorption per wavelength in SF_6 at 284·2 K, calculated by subtracting the classical absorption (equation 2.2.6) from the observed absorption (after Haebel, 1968).

relaxation equation (2.5.19) with the whole of the vibrational heat capacity relaxing. From the point of inflexion of the velocity dispersion curve or the peak of the absorption curve, the translational–vibrational relaxation time τ may be calculated by equations (2.5.17) and (2.5.19).

Such results suggest that the exchange of energy between all the molecular vibrations and translation takes place through a single vibrational mode which from the discussion of Section 4.2 is likely to be the lowest fundamental vibration frequency of the molecule, i.e. a series relaxation. This has been confirmed by vibrational fluorescence experiments in which the time required for intramolecular energy transfer has been determined from the phase lag in the re-emission of modulated infrared radiation. In CH_4, for example, the relaxation time for the removal of vibrational energy from ν_3 (90·6 THz) is $(7 \pm 1) \times 10^{-9}$ s which equals within experimental error the relaxation time for the intramolecular transfer of energy from ν_3 to the lowest fundamental ν_4 (39·2 THz) (Yardley and Moore, 1968): the vibrational–translational relaxation time in CH_4 is $2·0 \times 10^{-6}$ s. This intramolecular energy transfer occurs during an intermolecular collision so that the slight surplus of vibrational energy $(\nu_3 > 2\nu_4)$ can be transferred to translation. Hence the observed overall translational–vibrational relaxation time τ is longer than the relaxation time β of the lowest vibrational mode because of the extra energy which must pass through the lowest mode to excite the other vibrations:

$$\beta = C_1 \tau / C_i \tag{4.5.1}$$

where C_1 is the vibrational heat capacity of the lowest mode and C_i is the total vibrational heat capacity.

It is convenient to compare one substance with another in terms of the number of collisions which one molecule must undergo before energy transfer occurs. For a harmonic oscillator in which a number of vibrational levels are occupied, we may write (Herzfeld and Litovitz, 1959)

$$1/\beta = f'_{10} - f'_{01} \tag{4.5.2}$$

where f'_{10} and f'_{01} are the probabilities per molecule per second of collisional de-excitation and excitation of the lowest vibrational level ν_M of a polyatomic molecule. Equation (4.5.2) should be compared with the corresponding relation for a two-state system

$$1/\tau = k_f + k_b \tag{2.3.4}$$

Now at equilibrium

$$f'_{01}/f'_{10} = \overline{N'_1}/\overline{N'_0} = \exp(-h\nu_M/kT) \tag{4.5.3}$$

where $\overline{N'_0}$ and $\overline{N'_1}$ are the average number of molecules in the ground vibrational level and the first excited state of ν_M. Hence

$$1/f'_{10} = \beta[1 - \exp(-h\nu_M/kT)] \tag{4.5.4}$$

and the number of collisions Z'_{10} required to de-excite the lowest fundamental frequency by one quantum is

$$Z'_{10} = Z/f'_{10} = 1/p_{10} \tag{4.5.5}$$

where Z is the number of collisions which a molecule experiences in one second. Values of Z may be calculated from the kinetic theory of gases using values of the molecular collision diameter obtained from viscosity measurements.

In a small number of gases it is not possible to describe the vibrational relaxation by a single relaxation time. Figure 4.5.3 shows the dispersion of

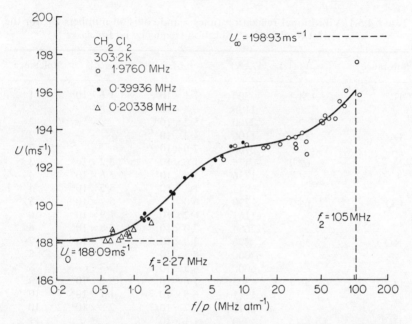

FIGURE 4.5.3. Dispersion of ultrasonic velocity in CH_2Cl_2 at 303·2 K using transducers of frequency 1·9760 MHz, o; 0·3994 MHz, •; 0·2034 MHz, △ (after Sette, Busala and Hubbard, 1955).

the ultrasonic velocity in CH_2Cl_2 at 303·2 K (Sette, Busala and Hubbard, 1955). The amplitude of the higher frequency relaxation process with an inflexion point at 105 MHz atm^{-1} corresponds to the relaxation of the heat capacity of the lowest vibration (8·5 THz) while the other vibrational modes, of which the lowest is 21·1 THz, give a single relaxation curve with an inflexion at 2·27 MHz atm^{-1}. Similar behaviour has been observed in ethane (Lambert and Salter, 1959), SO_2 (Lambert and Salter, 1957) and carbon suboxide (Hancock and Decius, 1969). In each case there is a substantial gap between the lowest fundamental frequency of vibration ν_M and the next ν_{M+1} such that $\nu_{M+1} > 2\nu_M$, and the relaxation time for intra-molecular vibrational energy transfer between ν_M and the higher modes is longer than the vibrational–translational relaxation time of ν_M. Such double relaxation is only detectable by ultrasonic techniques when the various vibrational levels make significant contributions ($>10\%$) to the vibrational heat capacity.

Some typical results for vibrational–translational energy transfer in pure gases are collected in Table 4.5.1. The low temperature results have been

TABLE 4.5.1. Vibrational relaxation times τ and collision numbers Z_{10} for the lowest fundamental vibration frequency ν_M for gases

Substance	ν_M (THz)	T (K)	τ (s)	Z'_{10}	Reference*
H_2	124·7	300	$1·1 \times 10^{-3}$	$1·6 \times 10^7$	1
		1100	$6·4 \times 10^{-6}$	—	2
		2700	$5·2 \times 10^{-7}$	—	2
D_2	89·6	1100	$1·2 \times 10^{-5}$	—	2
		2700	$7·6 \times 10^{-7}$	—	2
N_2	69·9	476	$1·6 \times 10^{-1}$	$8·2 \times 10^8$	3
		1736	$1·8 \times 10^{-3}$	$3·6 \times 10^6$	4
		6944	$1·3 \times 10^{-6}$	$5·5 \times 10^2$	4
CO	64·3	290	$6·0$	5×10^{10}	5
		1161	$6·2 \times 10^{-4}$	$1·9 \times 10^6$	6
		2654	$2·0 \times 10^{-5}$	$3·8 \times 10^4$	6
NO	57·1	296	$4·0 \times 10^{-7}$	$2·7 \times 10^3$	7
		2000	$5·3 \times 10^{-8}$	98	8
		6000	$4·5 \times 10^{-9}$	2·3	8
O_2	46·7	303	$1·6 \times 10^{-2}$	$9·8 \times 10^7$	9
		1111	$6·5 \times 10^{-5}$	$1·7 \times 10^5$	10
		5917	$2·7 \times 10^{-7}$	$2·8 \times 10^2$	10
HCl	86·5	290	$>1·1 \times 10^{-2}$	—	11
		1500	$3·5 \times 10^{-7}$	—	12
DCl	62·1	290	$1·0 \times 10^{-2}$	—	11
		1500	$6·0 \times 10^{-7}$	—	12
Cl_2	16·7	298	$4·9 \times 10^{-6}$	$4·4 \times 10^4$	13
		528	$1·6 \times 10^{-6}$	$6·8 \times 10^3$	13
HBr	76·7	290	$>1·5 \times 10^{-3}$	—	11
		1500	$5·0 \times 10^{-7}$	—	14
DBr	54·6	1500	$6·3 \times 10^{-7}$	—	15
HI	66·9	1500	$4·0 \times 10^{-7}$	—	14
DI	47·5	1500	$7·9 \times 10^{-7}$	—	15
Br_2	9·6	301	$8·5 \times 10^{-7}$	$4·4 \times 10^3$	13
		529	$4·6 \times 10^{-7}$	$1·3 \times 10^3$	13
I_2	6·4	385	$1·1 \times 10^{-7}$	$4·2 \times 10^2$	13
		526	$1·0 \times 10^{-7}$	$2·4 \times 10^2$	13
H_2O	47·8	410	$8·0 \times 10^{-9}$	76	16
D_2O	35·3	410	$7·0 \times 10^{-9}$	62	16
CO_2	20·2	293	$7·0 \times 10^{-6}$	$5·4 \times 10^4$	17

TABLE 4.5.1 (contd.)

Substance	ν_M (THz)	T (K)	τ (s)	Z'_{10}	Reference*
N_2O	17·7	300	$9·9 \times 10^{-7}$	$7·4 \times 10^3$	18
COS	15·8	293	$1·7 \times 10^{-6}$	$1·3 \times 10^4$	19
SO_2	15·6	293	$5·6 \times 10^{-8}$	$4·9 \times 10^2$	20
	34·5	293	$5·6 \times 10^{-7}$	$3·5 \times 10^3$	20
CS_2	11·9	273	$7·0 \times 10^{-7}$	$6·5 \times 10^3$	21
NH_3	28·5	298	$\leqslant 4·0 \times 10^{-10}$	$\leqslant 5$	22
ND_3	22·5	303	$1·3 \times 10^{-8}$	90	23
C_2H_2	18·4	298	$8·2 \times 10^{-8}$	$7·4 \times 10^2$	24
C_2D_2	15·3	298	$7·2 \times 10^{-8}$	$6·1 \times 10^2$	24
PH_3	29·7	303	$2·0 \times 10^{-7}$	$8·9 \times 10^2$	23
PD_3	21·8	303	$3·4 \times 10^{-7}$	$1·2 \times 10^3$	23
BF_3	14·4	298	$9·0 \times 10^{-8}$	$3·8 \times 10^2$	25
AsH_3	27·1	303	$3·1 \times 10^{-7}$	$3·6 \times 10^2$	26
AsD_3	19·8	303	$2·2 \times 10^{-7}$	$2·0 \times 10^2$	26
CH_4	39·2	298	$2·0 \times 10^{-6}$	$1·8 \times 10^4$	27
CD_4	29·9	298	$3·9 \times 10^{-6}$	$2·6 \times 10^4$	27
SiH_4	27·4	298	$1·1 \times 10^{-7}$	$\sim 6·0 \times 10^2$	27
SiD_4	20·2	298	$2·0 \times 10^{-7}$	$\sim 9·0 \times 10^2$	27
CH_3F	31·4	298	$2·7 \times 10^{-6}$	$6·8 \times 10^3$	24
CD_3F	29·7	298	$2·3 \times 10^{-6}$	$5·6 \times 10^3$	24
CH_3Cl	21·9	300	$2·0 \times 10^{-7}$	$1·0 \times 10^3$	28
CD_3Cl	21·0	294	$6·0 \times 10^{-7}$	$2·9 \times 10^3$	24
CH_2F_2	15·9	298	$4·0 \times 10^{-8}$	$2·1 \times 10^2$	29
CHF_3	15·2	298	$6·1 \times 10^{-7}$	$2·9 \times 10^3$	29
C_3O_2	1·9	300	$\leqslant 1·0 \times 10^{-9}$	$\leqslant 3$	30
	16·5	300	$4·8 \times 10^{-8}$	$4·6 \times 10^2$	30
CH_2Cl_2	8·5	303	$2·0 \times 10^{-9}$	18	31
	21·1	303	$9·5 \times 10^{-8}$	$3·6 \times 10^2$	31
CF_4	13·0	298	$8·2 \times 10^{-7}$	$2·1 \times 10^3$	29
CH_3Br	18·3	291	$7·2 \times 10^{-8}$	$3·3 \times 10^2$	32
CD_3Br	17·3	298	$1·4 \times 10^{-7}$	$2·2 \times 10^2$	33
$CHCl_3$	7·8	295	$1·5 \times 10^{-8}$	65	34
CH_3I	16·0	373	$4·5 \times 10^{-8}$	59	35
CCl_4	6·5	295	$1·9 \times 10^{-8}$	$1·31 \times 10^2$	36
$SiCl_4$	4·5	295	$5·9 \times 10^{-9}$	38	36
$GeCl_4$	4·0	295	$3·4 \times 10^{-9}$	17	36
$SnCl_4$	3·1	295	$2·5 \times 10^{-9}$	11	36
C_2H_4	24·3	298	$2·6 \times 10^{-7}$	$8·2 \times 10^2$	37
C_2D_4	17·5	298	$1·4 \times 10^{-7}$	$3·9 \times 10^2$	37
CH_3OH	10·1	305	$2·3 \times 10^{-9}$	10	38
SF_6	10·3	284	$7·6 \times 10^{-7}$	$2·1 \times 10^3$	39
C_2H_6	8·7	303	$7·1 \times 10^{-10}$	10	40
	24·6	303	$1·1 \times 10^{-8}$	92	40
C_3H_6	5·3	303	$1·5 \times 10^{-8}$	6	40

TABLE 4.5.1 (contd.)

Substance	ν_M (THz)	T (K)	τ (s)	Z'_{10}	Reference*
cyclo-C_3H_6	22·2	293	$3·1 \times 10^{-7}$	$1·3 \times 10^3$	32
C_3H_8	6·1	303	$1·6 \times 10^{-8}$	3	40
n-C_4H_{10}	3·0	298	$1·3 \times 10^{-9}$	1·6	41
cyclo-C_5H_{10}	6·2	298	$1·9 \times 10^{-9}$	2·6	41
neo-C_5H_{12}	6·0	298	$4·4 \times 10^{-9}$	4·2	41
C_6H_6	12·1	369	$1·5 \times 10^{-7}$	$3·0 \times 10^2$	42

*References
1. DeMartini and Ducuing (1966).
2. Kiefer and Lutz (1966).
3. Henderson (1962).
4. Millikan and White (1963).
5. Doyennette and Henry (1966).
6. Parker (1964).
7. Bauer and Sahm (1965).
8. Wray (1962).
9. Holmes, Smith and Tempest (1963).
10. Generalov (1963).
11. Ferguson and Read (1967).
12. Breshears and Bird (1969a).
13. Shields (1960).
14. Kiefer, Breshears and Bird (1969).
15. Breshears and Bird (1970).
16. Yamada and Fujii (1966).
17. Eucken and Nümann (1937).
18. Martinez, Strauch and Decius (1964).
19. Eucken and Aybar (1940).
20. Lambert and Salter (1957).
21. Gravitt (1960).
22. Jones and coworkers (1969).
23. Cottrell and Matheson (1963).
24. Stretton (1965).
25. Edmonds and Lamb (1958).
26. Cottrell and coworkers (1964).
27. Cottrell and Matheson (1962).
28. Amme and Legvold (1959).
29. Boade and Legvold (1965).
30. Hancock and Decius (1969).
31. Sette, Busala and Hubbard (1955).
32. Corran and coworkers (1958).
33. Lambert and Salter (1959).
34. Haebel (1965).
35. Fogg, Hanks and Lambert (1953).
36. Hinsch (1961).
37. Hudson, McCoubrey and Ubbelohde (1961).
38. Ener, Busala and Hubbard (1955).
39. Haebel (1968).
40. Holmes, Jones and Pusat (1964b).
41. Holmes, Jones and Lawrence (1966).
42. Lambert and Rowlinson (1951).

determined by the ultrasonic technique or the spectrophone. For comparison, some high temperature shock tube results have also been included. The substances in Table 4.5.1 are listed in order of increasing atomicity, while substances of a given atomicity are placed in order of increasing molecular weight. For each substance there is given the lowest fundamental vibration frequency ν_M, the overall vibrational relaxation time τ and Z'_{10}, the number of collisions required to remove one quantum from ν_M. As far as possible the most reliable results for a given compound have been chosen. Table 4.5.1 is far from being exhaustive, and more detailed compilations are given by Cottrell and McCoubrey (1961), Stevens (1967) and Gordon, Klemperer and Steinfeld (1968). Some results for the effect of additives on vibrational energy transfer are given in Table 4.5.2. The effect of adding a gas B to A is expressed as p_{10}^{AB}/p_{10}^{AA}, where p_{10}^{AB} is the probability of vibrational energy transfer in a collision between a molecule of A and one of B.

TABLE 4.5.2. Relative probabilities p_{10}^{AB}/p_{10}^{AA} for vibrational deactivation of a molecule A in the presence of an additive B

A	B	T (K)	p_{10}^{AB}/p_{10}^{AA}	Reference*
N_2	H_2O	560	1.1×10^3	1
CO	He	1000	60	2
	Ar	1000	0.1	3
	H_2	1000	5.0×10^2	3
	D_2	2000	3	4
	N_2	2000	30	4
	NO	2000	1.0×10^2	4
	O_2	2000	5	4
NO	N_2	300	1.1×10^{-3}	5
	CO	300	7.0×10^{-2}	5
	H_2O	300	20	5
	CO_2	300	0.48	5
O_2	He	295	3.5×10^2	6
	H_2	295	3.0×10^3	6
	D_2	295	3.0×10^2	6
	CH_4	295	1.3×10^4	7
	CD_4	295	3.3×10^3	7
Cl_2	HCl	500	58	8
	DCl	500	17	8
CO_2	He	303	11	9
	Ar	360	0.2	10
	H_2	320	1.8×10^2	11
	D_2	320	75	11
	H_2O	300	2.9×10^3	12
	D_2O	300	9.6×10^2	12
N_2O	He	293	5	13
	H_2	290	8	14
	D_2	292	13	14
	H_2O	294	71	15
	D_2O	294	30	15
SO_2	H_2O	298	13	16
	D_2O	298	38	16
CH_4	^{3}He	300	1.1	17
	^{4}He	300	0.74	17
	Ne	300	0.23	17
	Ar	300	0.13	17
	Kr	300	8.9×10^{-2}	17
	Xe	300	6.6×10^{-2}	17
	H_2	300	9.4	17
	HD	300	5.5	17
	D_2	300	1.7	17
	CO	300	0.18	17
	N_2	300	0.10	17
	O_2	300	0.30	17

TABLE 4.5.2 (contd.)

A	B	T (K)	p_{10}^{AB}/p_{10}^{AA}	Reference*
CH_3Cl	H_2O	298	26	18
	D_2O	298	27	18
	NH_3	298	12	19
	ND_3	298	12	19
CF_4	He	300	10	20
	Ne	300	2·5	20
	Ar	300	0·6	20
C_2H_4	H_2O	298	25	21
	D_2O	298	21	21
	NH_3	298	13	19
	ND_3	298	11	19
	C_2H_6	298	20	22
	C_3H_8	298	35	22
	$n\text{-}C_4H_{10}$	298	41	22
	$n\text{-}C_5H_{12}$	298	55	22
	$n\text{-}C_6H_{14}$	288	83	23

*References
1. Huber and Kantrowitz (1947).
2. Millikan (1964).
3. Hooker and Millikan (1963).
4. Sato, Tsuchiya and Kuratani (1969).
5. Basco, Callear and Norrish (1961).
6. Parker (1961).
7. Schnaus (1965).
8. Breshears and Bird (1969b).
9. Cottrell and Day (1966).
10. Simpson, Chandler and Strawson (1969).
11. Winter (1963).
12. Shields and Burks (1968).
13. Eucken and Nümann (1937).
14. Eucken and Jaacks (1935).
15. Wight (1956).
16. McCoubrey, Milward and Ubbelohde (1961a).
17. Yardley, Fertig and Moore (1970).
18. McCoubrey, Milward and Ubbelohde (1962).
19. Milward and Ubbelohde (1963).
20. Olson and Legvold (1963).
21. Hudson, McCoubrey and Ubbelohde (1961).
22. Arnold, McCoubrey and Ubbelohde (1958).
23. McCoubrey, Parke and Ubbelohde (1954).

Acoustic techniques are less useful in studying rotational–translational energy transfer (cf. Section 4.1). Figure 4.5.4 shows the ultrasonic absorption per wavelength μ in a mixture of 9·3% Ne in p-H_2 at 170 K: the maximum in the curve corresponds to the relaxation time for the transfer of energy between the $l = 0$ and $l = 2$ rotational levels of p-H_2. In gases other than hydrogen at low temperatures, a large number of rotational levels are populated and the observed relaxation time is an average for energy transfer to these states. Table 4.5.3 contains some typical values of the rotational relaxation time τ_r and the rotational collision number $Z_r(= Z\tau_r)$ for pure gases, while Table 4.5.4 illustrates the effect of additives on the rotational relaxation of hydrogen. The agreement between the ultrasonic results and those of other techniques is generally fair.

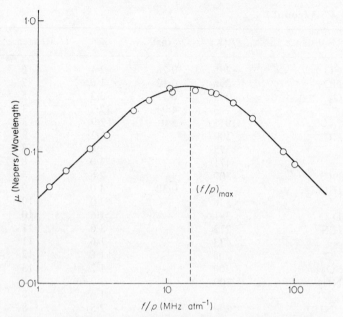

FIGURE 4.5.4. Ultrasonic absorption per wavelength in a mixture of 9·3% Ne in p-H_2 at 170 K, calculated by subtracting the classical absorption (equation 2.2.6) from the observed absorption (after Jonkman and coworkers, 1968a).

TABLE 4.5.3. Rotational relaxation times τ_r and collision numbers Z_r in gases

Substance	T (K)	τ_r (ns)	Z_r	Reference*
p-H_2 $(0 \rightleftharpoons 2)$	77	22	715	1
	170	19	411	1
	300	13	215	2
$(2 \rightleftharpoons 4)$	300	39	647	2
o-H_2 $(1 \rightleftharpoons 3)$	300	23	371	2
$(3 \rightleftharpoons 5)$	300	89	1460	2
n-H_2	873	43	303	3
	1073	50	307	3
o-D_2	31	10	370	1
	91	13	268	1
n-D_2	773	45	225	3
	1073	75	328	3
HF	374	—	9·5	4
DF	374	—	4·1	4
N_2	295	0·76	5·5	5
	773	3·0	11·3	3
	1073	5·0	15·4	3

TABLE 4.5.3 (contd.)

Substance	T (K)	τ_r (ns)	Z_r	Reference*
CO	$\sim$298	0·28	2·1	6
NO	296	0·12	1·0	7
O_2	295	0·51	3·3	5
	773	4·0	12·9	3
	1073	5·0	12·9	3
HCl	300	—	6·2	8
	375	—	3·6	8
	471	—	3·0	8
DCl	300	—	2·6	8
H_2O	323	0·32	4·0	9
	526	—	2·0	10
D_2O	381	—	1·6	10
H_2S	273	—	3·0	11
	473	—	1·6	11
CO_2	298	—	1·6	12
SO_2	300	—	1·2	13
NH_3	300	—	2·3	8
	475	—	1·6	8
$^{15}NH_3$	300	—	2·5	8
ND_3	300	—	1·6	8
C_2H_2	287	—	1·8	14
CH_4	303	1·1	12	15
	1073	7·0	30	16
CD_4	303	0·73	7	15
CH_3F	300	—	2·2	17
CH_2F_2	300	—	1·2	17
CHF_3	300	—	1·6	17
CF_4	288	—	3·0	14
$CHCl_3$	295	—	2–3	18
C_2H_4	283	—	2·4	14
	773	3·0	17	16
SF_6	284	0·6	5	19
C_2H_6	286	—	4·0	14
cyclo-C_3H_6	303	0·36	5	15
neo-C_4H_{10}	298	0·2	3·7	20
cyclo-C_5H_{10}	298	0·1	2·4	20

*References

1. Jonkman and coworkers (1968a, b).
2. Valley and Amme (1969).
3. Winter and Hill (1967).
4. Baker (1967).
5. Fujii, Lindsay and Urushihara (1963).
6. Bauer and Kosche (1966).
7. Bauer and Sahm (1965).
8. Baker and Brokaw (1965).
9. Roesler and Sahm (1965).
10. Baker and Brokaw (1964).
11. Barua, Manna and Mukhopadhyay (1968).
12. Holmes, Jones and Lawrence (1964).
13. Baker and de Haas (1964).
14. O'Neal and Brokaw (1963).
15. Holmes, Jones and Pusat (1964a).
16. Hill and Winter (1968).
17. Crapo and Flynn (1965).
18. Haebel (1965).
19. Haebel (1968).
20. Holmes, Jones and Lawrence (1966).

TABLE 4.5.4. Relative probabilities p_{10}^{AB}/p_{10}^{AA} for rotational energy transfer from a molecule A in the presence of an additive B (Jonkman and coworkers, 1968a, b)

A	B	T (K)	p_{10}^{AB}/p_{10}^{AA}
p-H_2	o-H_2	77·0	1·8
	³He	170	2·1
	⁴He	170	2·0
	Ne	170	2·9
	Ar	170	2·1
	Kr	170	1·6
	Xe	170	1·2
o-D_2	⁴He	90·5	1·8
	Ne	90·5	2·9
	Ar	90·5	2·8

4.6 Vibrational–Translational Energy Transfer: Comparison of Theory with Experiment

Table 4.5.1 shows that a large number of molecular collisions are required before vibrational–translational energy transfer occurs. The qualitative considerations of Section 4.1 correctly predict the general trend of the experimental results: the lower the fundamental vibration frequency ν_M the smaller the vibrational relaxation time, while the higher the temperature and hence the translational velocity the smaller the relaxation time.

The more rigorous molecular theory of vibrational energy transfer outlined in Section 4.2 may conveniently be tested by studying the temperature dependence of the vibrational relaxation time. For a given substance the vibration frequencies, intermolecular forces and molecular masses are independent of temperature. Equation (4.2.17) thus suggests that a plot of log Z_{10} against $T^{-\frac{1}{3}}$ should be approximately linear if the pre-exponential factor in the equation and attractive forces are negligible. The available results for oxygen (White and Millikan, 1963) are plotted in Figure 4.6.1. As in many other gases, a good straight line is obtained.

From the slope of this line one may calculate a value of α, the constant of the intermolecular potential in equation (4.2.8). The values of α obtained in this way are substantially higher than those determined from viscosity measurements (Arnold, McCoubrey and Ubbelohde, 1958). Considerably improved agreement is obtained by following the suggestion of Zener (1931) that the relative translational velocity in an inelastic collision should be symmetrized, that is the mean of the approach and the recession velocity should be taken. A further improvement is obtained by including the attractive part of the intermolecular potential in equation (4.2.8). The values of α obtained under these conditions are still typically some 30% higher than those from viscosity (McCoubrey, Milward and Ubbelohde, 1961b). Such a discrepancy is not wholly unexpected. Although all collisions are effective in momentum transfer and viscosity, only high energy

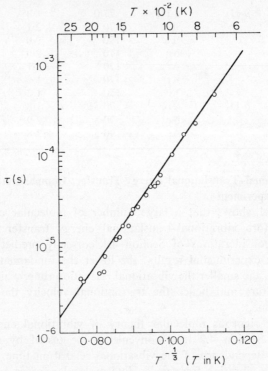

FIGURE 4.6.1. Vibrational relaxation time of oxygen as a function of $T^{-\frac{1}{3}}$ (after White and Millikan, 1963).

collisions lead to vibrational energy transfer. Thus an arbitrary function such as the Lennard–Jones potential with constants derived from viscosity values would not be expected to give an accurate estimate of the probability of energy transfer. Moreover such a centro-symmetric function may not be strictly applicable at the short molecular separations involved in energy transfer.

If the effective value of α can be estimated (Herzfeld and Litovitz, 1959), then it is possible to make an absolute calculation of the probability of energy transfer in a collision. Such calculations overestimate the transition probability, the discrepancy being much greater for diatomic molecules such as nitrogen than for spherical molecules such as methane. The theoretical treatment neglects the fact that not all collisions occur with molecular orientations favourable to energy transfer. This 'steric factor' is likely to be closer to unity in methane than in nitrogen and hence the agreement of theory with experiment is better in methane. Thus the quantitative calculation of vibrational relaxation times is limited by the uncertainty in the values of α and the steric factor. Despite these limitations the calculated probabilities of energy transfer are always within a factor of ten of the observed probabilities although the predicted dependence on temperature is somewhat too steep.

Many refinements of the molecular theory have been made. These have included the use of different potential functions (Cottrell and Ream, 1955; Calvert and Amme, 1966) including angle dependent functions (Parker, 1964), dipolar interactions (Tanczos, 1956) and attractive potentials (Blythe, Cottrell and Read, 1961). Steric factors have also been calculated explicitly for a number of systems and found to depend upon temperature as well as on the molecular species present (Shin, 1967a, b). Although such modifications produce a useful improvement in the agreement between theory and experiment, the limitations of the molecular theory have also become apparent. For example, Rapp and Sharp (1963) indicate that the perturbation approximation is not always adequate for the high energy collisions involved in energy transfer, while Mies' (1965) calculations using an exact $He–H_2$ potential energy surface have shown the limitations of the pairwise, centro-symmetric intermolecular potential calculated from viscosity. Nevertheless the molecular theory with realistic modifications can calculate vibrational relaxation times to within a factor of three.

The effect of additives on the efficiency of vibrational energy transfer is complicated by rotational–vibrational energy transfer (Section 4.7) and the resonance transfer of vibrational energy (Section 4.8). Where such effects are absent, the probability of energy transfer in a heteromolecular collision may be calculated in the same way as for a homomolecular collision. As in pure gases, the calculations are limited chiefly by uncertainties about the precise form of the intermolecular potential energy function. The general trends shown in Table 4.5.2 are well reproduced, such as the high efficiency of helium with respect to argon in vibrational energy transfer.

4.7 Vibrational–Rotational Energy Transfer

In the pairs of isotopically substituted compounds CH_4 and CD_4, SiH_4 and SiD_4, and PH_3 and PD_3, the vibrational relaxation time of a hydride is longer than that of the corresponding deuteride despite the higher vibrational frequencies of the hydrides. Cottrell and Matheson (1962) ascribed this to the sequence

$$\text{vibration} \rightleftharpoons \text{rotation} \rightleftharpoons \text{translation}$$

in molecules of low moment of inertia where the peripheral velocity of rotation is greater than the translational velocity. The simple theory outlined in Section 4.3 adequately reproduces the ratio of relaxation times in such isotopically substituted compounds, although the absolute values of the calculated relaxation times are substantially shorter than those observed. Confirmation of the role of rotation in vibrational energy transfer in these compounds comes from a comparison of the relaxation times of PH_3 and AsH_3 (Cottrell and coworkers, 1964): these molecules have similar lowest vibration frequencies, moments of inertia and rotational velocities but differ in molecular weight and hence in translational velocity. Vibrational–translational energy transfer theory predicts that the ratio $\beta(AsH_3)/\beta(PH_3)$ should be 150 at room temperature, whereas the vibrational-rotational theory gives a ratio of unity. Experimentally $\beta(AsH_3)/\beta(PH_3)$ equals 1·5. Vibrational–rotational energy transfer would also be expected in NH_3 and ND_3: in this case, however, the extremely short vibrational relaxation time of NH_3 has been ascribed to the impulsive intermolecular force arising from the inversion of the NH_3 molecule during a collision (Cottrell and Matheson, 1963).

In all these molecules the lowest fundamental vibration frequency is a bond bending which could be expected to interact with a rotating molecule particularly easily (Figure 4.4.1). But in the hydrogen halides the same phenomenon is observed, the vibrational relaxation time of deuterium iodide, for example, being substantially longer than that of hydrogen iodide (Breshears and Bird, 1970). In methyl chloride and methyl bromide the lowest mode is the carbon–halogen stretching frequency ν_3. This mode involves much less atomic movement than the next mode ν_6, however, and Stretton (1965) has shown that energy transfer will occur preferentially through ν_6. Since ν_6 involves substantial bond bending, vibrational–rotational energy transfer is facilitated, and thus the vibrational relaxation times of the deuterated methyl halides are longer than those of the corresponding hydrogenated methyl halides.

Although the vibrational relaxation time of C_2H_4 is longer than that of C_2D_4, the ratio $\beta(C_2D_4)/\beta(C_2H_4)$ is significantly larger than predicted for a direct translation–vibration process (Hudson, McCoubrey and Ubbelohde, 1961). In the acetylenes, however, the moments of inertia are larger and the experimental ratio $\beta(C_2D_2)/\beta(C_2H_2)$ is close to the ratio calculated from translation–vibration theory (Stretton, 1965). Thus it appears that vibrational–rotational energy transfer predominates in molecules where the rotational velocity is substantially greater than the translational velocity. This offers an explanation of the division of a plot of $\log Z'_{10}$ versus ν_M into two distinct groups of compounds (Lambert and Salter, 1959) as shown in Figure 4.7.1. The group with high Z'_{10} represents

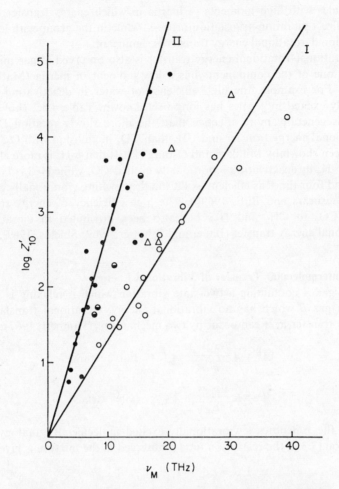

FIGURE 4.7.1. The Lambert–Salter plot showing the relation between the collision number Z'_{10} and the lowest fundamental vibration frequency ν_M. Values listed in order from the bottom upwards are: ●, II, molecules with no hydrogen atoms: $SnCl_4$, C_2F_4, $GeCl_4$, $SiCl_4$, CBr_2F_2, $CClBrF_2$, CCl_4, CCl_2F_2, CCl_3F, I_2, BF_3, $CBrF_3$, SO_2, $CClF_3$, SF_6, CF_4, Br_2, CS_2, COS, N_2O, Cl_2, CO_2; ◒, molecules containing one hydrogen atom: $CHCl_3$, $CHCl_2F$, $CHClF_2$, CF_3H; △, molecules containing two or more deuterium atoms: C_2D_2, C_2D_4, SiD_4, CD_4; ○, I, molecules containing two or more hydrogen atoms: CH_2Cl_2, CH_2CHCl, CH_2CHBr, CH_2FCl, CH_3I, CH_2CHF, CH_2CF_2, CH_2F_2, CH_3Br, C_2H_2, CH_3Cl, cyclo-C_3H_6, CH_3F, SiH_4, CH_4.

molecules with high moments of inertia in which energy transfer occurs by a direct vibration–translation process, while in the group with low Z'_{10} vibrational–rotational energy transfer is significant.

Vibrational–rotational energy transfer is also observed in gas mixtures where one of the components has a low moment of inertia (Matheson, 1962). For example, the high efficiency of water in deactivating vibrationally excited molecules has long been known. Table 4.5.2 shows that H_2O is generally more effective than the more slowly rotating D_2O in vibrational energy transfer and CH_4 than CD_4 in collision with O_2. It has also been shown by Millikan and Osburg (1964) that p-H_2 is more efficient than o-H_2 in deactivating vibrationally excited CO, while HCl is between two and four times as efficient as DCl in deactivating vibrationally excited Cl_2 (Breshears and Bird, 1969b). The high efficiency of energy transfer from CO_2 to CH_4 and H_2S has also been attributed to vibrational–rotational energy transfer (Bauer and Schotter, 1969; Shields, 1969).

4.8 Intermolecular Transfer of Vibrational Energy

If a gas A containing a two-state vibration which is relaxing is mixed with a gas B which has no vibrational energy, vibrational–translational energy transfer in A can occur by two mechanisms (Lambert, 1967):

(1)
$$A^* + A \underset{k_{01}{}^{AA}}{\overset{k_{10}{}^{AA}}{\rightleftharpoons}} A + A \quad \text{(vib} \rightleftharpoons \text{trans)}$$

(2)
$$A^* + B \underset{k_{01}{}^{AB}}{\overset{k_{10}{}^{AB}}{\rightleftharpoons}} A + B \quad \text{(vib} \rightleftharpoons \text{trans)}$$

where the * denotes a vibrationally excited molecule. By analogy with equation (2.3.4) the relaxation time β observed in the mixture is given by

$$1/\beta = k_{01}^{AA} + k_{01}^{AB} + k_{10}^{AA} + k_{10}^{AB} \tag{4.8 1}$$

where k_{01}^{AA} is the average number of 0–1 vibrational transitions in A–A collisions per molecule per second at 1 atm, and the other symbols have a similar meaning. In such a mixture the value of $k_{01}^{AA} + k_{10}^{AA}$ decreases as x_A, the mole fraction of A, decreases, giving where A and B are similar in size and shape

$$k_{01}^{AA} + k_{10}^{AA} = x_A/\beta_{AA} = (1 - x_B)/\beta_{AA} \tag{4.8.2}$$

where β_{AA} is the vibrational relaxation time of pure A at 1 atm. Likewise

$$k_{01}^{AB} + k_{10}^{AB} = x_B/\beta_{AB} \tag{4.8.3}$$

where β_{AB} is the relaxation time of a hypothetical mixture in which only

AB collisions cause transitions. Hence

$$\frac{1}{\beta} = \frac{1 - x_B}{\beta_{AA}} + \frac{x_B}{\beta_{AB}} \qquad (4.8.4)$$

and a plot of $1/\beta$ against x_B should be a straight line.

If both A and B have vibrational energy, there are now five possible energy transfer processes:

(1) $\qquad\qquad A^* + A \rightleftharpoons A + A \qquad$ (vib $\rightleftharpoons$ trans)

(2) $\qquad\qquad A^* + B \rightleftharpoons A + B \qquad$ (vib $\rightleftharpoons$ trans)

(3) $\qquad\qquad A^* + B \rightleftharpoons A + B^* \qquad$ (vib $\rightleftharpoons$ vib $\pm$ trans)

(4) $\qquad\qquad B^* + B \rightleftharpoons B + B \qquad$ (vib $\rightleftharpoons$ trans)

(5) $\qquad\qquad B^* + A \rightleftharpoons B + A \qquad$ (vib $\rightleftharpoons$ trans)

In process (3), vibrational energy from a vibrationally excited A molecule is transferred in a collision to vibration of a B molecule: if the vibrational quanta in the two molecules are not equal, the energy difference is equilibrated with translation.

The mechanism of vibrational energy transfer in a mixture of two relaxing gases depends on the relative rates of processes (1) to (5). If (3) is very slow, then A and B will relax separately: the mixture will show double dispersion of the ultrasonic velocity and each of the relaxation times will vary with the composition of the mixture according to equation (4.8.4). On the other hand, if (3) is very rapid and vibrational energy is easily exchanged between A and B, the vibrational energy of both species will equilibrate with translation by a common mechanism with a single relaxation time. If, as is likely, this energy transfer is homomolecular, i.e. (1) or (4), the rate is proportional to x^2 and there should be a near quadratic dependence of $1/\beta$ on x^2. Such behaviour has been observed by Lambert, Parks-Smith and Stretton (1964) in a mixture of SF_6 ($\nu_M = 10 \cdot 3$ THz) and $CHClF_2$ ($\nu_M = 11 \cdot 1$ THz). All concentrations in the mixture have a single relaxation time, the reciprocal of which varies as the square of the mole fraction of $CHClF_2$ (Figure 4.8.1). The collision numbers for the pure gases at 300 K are $Z_{SF_6-SF_6} = 1005$ and $Z_{CHClF_2-CHClF_2} = 122$. From the shape of the curve in Figure 4.8.1, it is possible to calculate $Z_{SF_6-CHClF_2} = 50$.

A further possibility is that reaction (3) is faster than (1) and (2) but slower than (4) and (5). B will now relax independently by (4) or (5) giving a linear dependence of $1/\beta_B$ with x. A will relax by (3) followed by (4) or (5) with (3) the rate-determining step so that $1/\beta_A$ is also linear with x. The expected double dispersion has been observed by Lambert and coworkers (1962) in a number of mixtures. Figure 4.8.2 shows the expected

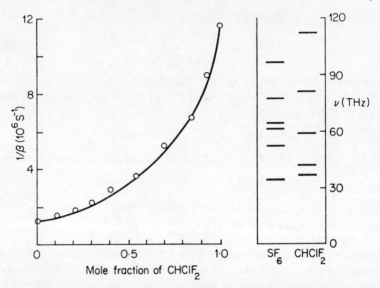

FIGURE 4.8.1. Reciprocal relaxation times and energy level diagram for $SF_6 + CHClF_2$ mixtures. ○, observed points; —, curve calculated assuming $1/\beta \propto x^2$, where x is the mole fraction of $CHClF_2$ (after Lambert, Parks-Smith and Stretton, 1964).

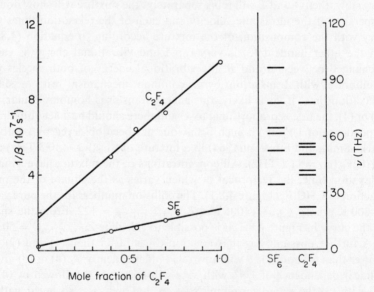

FIGURE 4.8.2. Reciprocal relaxation times and energy level diagram for $SF_6 + C_2F_4$ mixtures. ○, observed points (after Lambert and coworkers, 1962).

linear variation of both reciprocal relaxation times with concentration in a mixture of SF_6 and C_2F_4. There is near resonance between the 10·3 THz fundamental of SF_6 and the first harmonic of the 5·7 THz mode of C_2F_4. $Z_{SF_6-SF_6}$ at 300 K is 1005, $Z_{C_2F_4-C_2F_4}$ is 5·5, while $Z_{SF_6-C_2F_4}$ is 70.

Table 4.8.1 gives some typical results for vibrational–vibrational energy transfer in gases. The number of collisions required for intermolecular transfer of vibrational energy is typically 50, with collisions requiring

TABLE 4.8.1. Collision numbers at 300 K for the transfer of one quantum from vibration ν_A of molecule A into i quanta of vibration ν_B of molecule B (Lambert, Parks-Smith and Stretton, 1967)

A	B	ν_A (THz)	ν_B (THz)	i	$\Delta\nu$ (THz)	Z_{AB}	Z_{AA}	Z_{BB}
		Singly-relaxing mixtures						
SF_6	$CHClF_2$	10·3	11·1	1	+0·8	50	1005	122
C_2H_4	C_2H_6	24·3	24·6	1	+0·3	40	970	74
		Doubly-relaxing mixtures						
CCl_2F_2	CH_3OCH_3	7·8	7·5	1	−0·3	5	73	<3
CH_3Cl	CH_3OCH_3	21·9	7·5	3	+0·6	70	421	<3
SF_6	CH_3OCH_3	10·3	4·9	2	−0·5	80	1005	<3
CHF_3	C_2F_4	15·2	15·2	1	0·0	50	1500	5·5
SF_6	C_2F_4	10·3	5·7	2	+1·1	70	1005	5·5
CF_4	C_2F_4	13·0	6·6	2	+0·2	110	2330	5·5

multiple quantum transfers being less efficient than single quantum transfers. The high efficiency of this process emphasizes the need for high purity when studying vibrational relaxation in substances with long vibrational relaxation times: the presence of a trace of a high molecular weight compound which has a short relaxation time and many vibrational frequencies available for intermolecular vibrational energy transfer can drastically shorten the vibrational relaxation time of the substance being studied.

Vibrational–vibrational energy transfer is also observed in diatomic molecules. For example, the vibrational relaxation time of nitrogen is greatly shortened by the addition of carbon dioxide on account of the transfer of energy from the 69·9 THz vibration of nitrogen to the 70·4 THz vibration of CO_2 (Henderson and coworkers, 1969a; Taylor and Bitterman, 1969). A similar effect is observed on addition of CD_4 to nitrogen (Henderson, Burbank and Glatzel, 1969b). Both the classical and quantum mechanical theories of energy transfer (Section 4.2) have been extended to such cases, and the agreement between theory and experiment is good (Sharma and Brau, 1969; Berend and Benson, 1969).

4.9 Rotational–Translational Energy Transfer: Comparison of Theory with Experiment

If hydrogen is excluded, there are two significant features to be seen in the results for rotational–translational energy transfer in Table 4.5.3. Firstly, rotational energy transfer is very efficient, the collision number Z_r normally being less than 10. Secondly, Z_r has a negative temperature dependence, the collision number increasing with increasing temperature: this unusual behaviour is in contrast to the decrease in the vibrational collision number with increasing temperature.

The first of these features can be qualitatively predicted by any of the classical approaches outlined in Section 4.3. The predicted collision numbers are within a factor of two of those observed experimentally, although the predicted values are usually too high. Agreement with experiment is improved by including an attractive intermolecular potential and considering multiple 'chattering' collisions in non-spherical molecules. The decrease in Z_r with increasing dipole moment predicted by Zeleznik (1967) is also observed.

The negative temperature dependence is less readily predicted. The observed dependence is much steeper than is predicted by Parker's (1959) classical theory. The most successful approach is the multilevel theory of Raff and Winter (1968). This contains one adjustable parameter which is fitted to the experimental results at one temperature. There is then excellent agreement with experiment in H_2, D_2 and N_2 over a 700 K temperature range.

Both p-H_2 below 170 K and o-D_2 below 90 K show the expected decrease in rotational collision number with increasing temperature. In this temperature range only the two lowest rotational energy levels are populated. Jonkman and coworkers (1968c) have shown that the available theoretical treatments do not give satisfactory agreement with experiment, particularly when applied to the results for mixtures in Table 4.5.4. At higher temperatures, however, where many rotational levels are populated, the dependence upon temperature of Z_r is reversed and the Raff and Winter (1968) treatment may be applied.

It is clear that techniques based on ultrasonic propagation or thermal conductivity are of limited value in studying translational–rotational energy transfer in gases. Only when the relaxation times of individual rotational levels are available will a proper comparison of theory with experiment be possible. A useful technique is that of microwave double resonance which has been applied by Unland and Flygare (1966) to study the rate of the $J = 1 \rightarrow 2$ transition in COS. Specific rotational transitions can also be studied with molecular beams (Blythe, Grosser and Bernstein, 1964).

CHAPTER 5

VISCOELASTICITY

5.1 Viscoelasticity of a Medium

The classical theories of elasticity and hydrodynamics classify deformable substances into the idealized categories of perfectly elastic solids and perfectly viscous fluids according to their response to the application of a stress. A perfectly elastic solid is one which recovers its original size and shape after being subjected to a deformation: during this deformation, the solid obeys Hooke's law which states that the deformation or strain produced by an applied force is proportional to the magnitude of the applied force or stress.

Figure 5.1.1 is a cross-section ABCD of a cube of material subjected to tangential forces which cause a change in the shape of the cube, the face

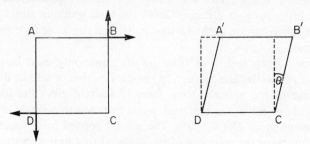

FIGURE 5.1.1. Shear strain.

ABCD becoming the rhombus A'B'CD. For small strains the volume of the cube and the size of the face do not change, since the area ABCD equals the area A'B'CD. This type of strain is called shear strain, and it is measured by the angular deformation θ. Hooke's law for a body experiencing a shear stress may be written

$$\sigma = G\gamma \tag{5.1.1}$$

where σ is the shear stress or tangential force per unit area, γ is the shear strain θ, and the constant of proportionality G is the modulus of rigidity of the material.

Application of a shear stress to a perfectly viscous liquid causes the liquid to flow in order to relieve the stress. The relative motion of different layers of the flowing liquid leads to a resistance to flow and energy dissipation, and the liquid is said to exhibit viscosity. At sufficiently low shear stresses liquids obey Newton's law

$$\sigma = \eta_s\, \partial\gamma/\partial t \qquad (5.1.2)$$

where η_s is the coefficient of shear viscosity: the stress is directly proportional to the rate of strain but independent of the magnitude of the strain. Such a liquid is said to be Newtonian.

Real materials may show deviations from ideal Hookean and Newtonian behaviour. Time anomalies may occur in which the stress depends on both the strain and the rate of strain together, and also on higher time derivatives of the strain. A material showing such time anomalies combines properties characteristic of liquids and solids and is said to be viscoelastic. Stress anomalies are also found in which the strain in a solid or the rate of strain in a liquid is not directly proportional to the stress. We shall not consider such non-linear behaviour, and shall concentrate on linear viscoelastic materials in which only time anomalies occur: the ratio of stress to strain is a function of time only and not of the amplitude of the stress. Moreover, a linear viscoelastic material obeys the Boltzmann (1876) superposition principle. This states that if a number of different stresses are applied to a body at different times, the total resultant strain is given by a linear superposition of the various individual strains at their respective elapsed times.

Two types of applied force function are commonly used in studying viscoelastic materials. The first is a harmonic function of infinite duration: the response of the material to a sinusoidal strain gives the (complex) rigidity modulus G^* of the material, while response to a sinusoidal stress gives the compliance $J^* (= 1/G^*)$. The second applied force function is a unit step function: the response of the material to a unit step strain gives the stress relaxation function $\psi(t)$, while response to a unit step stress gives the creep function $J(t)$. G^* and J^* are functions of frequency, while the relaxation and creep functions are functions of time. We shall consider primarily the measurement and interpretation of the response of a material to harmonic functions, although the relations to step functions will be discussed where appropriate.

5.2   Molecular Picture of Viscoelastic Relaxation

Consider a liquid confined between two parallel plates, one of which oscillates sinusoidally along its length while the other is fixed. For low frequencies of oscillation the liquid behaves as a liquid: it flows in an oscillatory fashion in response to the oscillatory shear strain. All of the driving energy is dissipated in viscous flow of the liquid and no energy is stored elastically. The period of the oscillation is much greater than the time between diffusive jumps of the liquid molecules so that liquid flow occurs as a small directional drift superimposed on the random, thermal, molecular motion.

Suppose that the frequency of the oscillation is now increased so that its period is much shorter than the diffusional jump-time of molecules in the liquid. Since no molecular diffusion occurs during the period of shear strain, no energy is dissipated in viscous flow, that is, the viscosity is zero. Instead the energy is stored elastically, and the 'liquid' is behaving on this time-scale as an amorphous 'solid' with a rigidity modulus comparable to that of a normal crystalline solid.

At some intermediate frequency where the period of the applied oscillatory strain is comparable to the diffusive jump-time, the liquid will show both viscous and elastic properties. The change from viscous to elastic behaviour with increasing frequency is called viscoelastic relaxation, and the diffusional jump-time of the liquid molecules is approximately equal to the viscoelastic relaxation time. The object of studies of viscoelastic relaxation using alternating shear strains (i.e. shear waves) is to understand the molecular parameters which determine the viscoelastic relaxation time and rigidity modulus of a liquid, and hence to obtain a better understanding of liquid viscosity and other transport properties.

Well-known examples of viscoelastic liquids are tar and pitch. When subjected to slow shear strain (< 1 Hz, say) these liquids flow, but when subjected to an instantaneous strain as from a blow from a hammer they shatter in a typical solid fashion. Provided that the time-scale of the observation is sufficiently short, all liquids show viscoelastic relaxation, although in low viscosity liquids such as argon the viscoelastic relaxation time may be as short as 10^{-13} s.

The discussion of Section 5.1 suggested that when a body is subjected to a shear strain, no volume change occurs: hence no temperature change would be expected. This has been confirmed experimentally in metals by Muller (1969). When a shear wave is applied to a viscoelastic medium there should be no simultaneous temperature variation: the temperature rise in the medium resulting from viscous flow appears as a random increase in the translational energy of the molecules of the medium and not as a temperature variation in phase with the applied strain. This is in marked contrast to the situation with a longitudinal wave. Thus temperature-sensitive equilibria in a medium will not be affected by a shear wave so that the latter can be used to study molecular diffusional motion free from the complications of other relaxation processes in the medium.

5.3 Propagation of a Shear Wave in a Viscoelastic Medium

Consider the propagation of a transverse shear wave along the z direction of a viscoelastic medium. Let σ_{xz} be the shear stress acting in the

4

x direction, and ξ the displacement of the medium. Equation (5.1.1) may then be written

$$\sigma_{xz} = G^*\gamma = G^* \, \partial\xi/\partial z \qquad (5.3.1)$$

Equating the driving force on an element of the medium of unit volume to the product of its mass and acceleration, we have

$$\rho\frac{\partial^2\xi}{\partial t^2} = \frac{\partial\sigma_{xz}}{\partial z} \qquad (5.3.2)$$

$$= G^*\frac{\partial^2\xi}{\partial z^2} \qquad (5.3.3)$$

by (5.3.1). This is the equation describing the propagation of a shear wave in a viscoelastic medium. For a progressive shear wave, the solution is the same as equation (2.2.3):

$$\xi = \xi_0 \exp\left(-\alpha z\right)\exp\left[i\omega(t-z/U_s)\right] \qquad (5.3.4)$$

where α is the absorption coefficient of the shear wave per metre of path length in the medium, U_s is the real phase velocity, ω is the angular frequency and ξ_0 is the initial value of ξ. Equation (5.3.4) may also be written

$$\xi = \xi_0 \exp\left[i\omega\left\{t-z\left(\frac{1}{U_s}+\frac{\alpha}{i\omega}\right)\right\}\right]$$

$$= \xi_0 \exp\left[i\omega(t-z/U_s^*)\right] \qquad (5.3.5)$$

where the complex velocity U_s^* is

$$\frac{1}{U_s^*} = \frac{1}{U_s}-\frac{i\alpha}{\omega} \qquad (5.3.6)$$

and also

$$U_s^* = (G^*/\rho)^{\frac{1}{2}} \qquad (5.3.7)$$

The rigidity modulus G^* is frequency dependent and complex:

$$G^* = G'+iG'' \qquad (5.3.8)$$

G' is an elastic or storage modulus in phase with the applied shear strain and G'' is a loss modulus, 90° out of phase with the applied strain.

We can define a complex viscosity η^*:

$$\eta^* = \eta'-i\eta'' = G^*/i\omega \qquad (5.3.9)$$

Hence

$$\eta' = G''/\omega \quad \text{and} \quad \eta'' = G'/\omega \qquad (5.3.10)$$

We can also define a complex compliance J^*:

$$\gamma/\sigma_{xz} = J^* = J'-iJ'' = 1/G^* \qquad (5.3.11)$$

Hence

$$J' = \frac{G'}{(G'^2 + G''^2)}; \quad J'' = \frac{G''}{(G'^2 + G''^2)} \tag{5.3.12}$$

The quantity which is usually measured experimentally is the complex shear mechanical impedance of the medium Z_L which is defined as the ratio of the shear stress to the particle velocity:

$$Z_L = R_L + iX_L = -\sigma_{xz}/(\partial \xi/\partial t) \tag{5.3.13}$$

Substitution of the value of σ_{xz} from (5.3.1) gives

$$Z_L = -G* \frac{(\partial \xi/\partial z)}{(\partial \xi/\partial t)} \tag{5.3.14}$$

This equation may be simplified by taking the partial derivatives of equation (5.3.5) to give

$$Z_L = G*/U_s^* \tag{5.3.15}$$

Substituting the value of U_s^* in (5.3.7) we obtain

$$Z_L = (\rho G*)^{\frac{1}{2}} \tag{5.3.16}$$

or

$$(R_L + iX_L) = [\rho(G' + iG'')]^{\frac{1}{2}}$$

Hence

$$G' = (R_L^2 - X_L^2)/\rho; \quad G'' = 2R_L X_L/\rho \tag{5.3.17}$$

This equation relates the measured quantities R_L and X_L to the components of the rigidity modulus $G*$.

The values of the various viscoelastic parameters in a Newtonian liquid and a Hookean solid may be derived from the general equations given above. In a Newtonian liquid where $G' = 0$ by definition, $\eta'' = 0$ and $J' = 0$. G'' is then related to the static, steady flow viscosity η_s

$$G'' = \omega \eta_s \tag{5.3.18}$$

From (5.3.17), $R_L = X_L$, and thus

$$R_L = X_L = (\pi f \eta_s \rho)^{\frac{1}{2}} \tag{5.3.19}$$

$f = \omega/2\pi$ being the frequency. These values of R_L, X_L, G' and G'' can be substituted into (5.3.15) to give $U*$, and equation (5.3.6) used to calculate the phase velocity and absorption coefficient of shear waves in a Newtonian liquid:

$$U_s = (2\omega \eta_s/\rho)^{\frac{1}{2}}; \quad \alpha = (\omega \rho/2\eta_s)^{\frac{1}{2}} \tag{5.3.20}$$

The shear velocity is frequency dependent, and the absorption is extremely large. Take water at room temperature, for example, where $\rho \approx 10^3$ kg m^{-3} and $\eta \approx 10^{-3}$ N s m^{-2} at a frequency f of 1 MHz,

$$U_s = 3 \cdot 5 \text{ ms}^{-1} \quad \text{and} \quad \alpha = 1 \cdot 8 \times 10^6 \text{ nepers m}^{-1}$$

In a perfect Hookean solid where $\eta' = 0$ by definition, $G'' = 0$ and $J'' = 0$. G' equals the rigidity modulus of the solid G, while from (5.3.17)

$$R_S = (\rho G)^{\frac{1}{2}}; \quad X_S = 0 \tag{5.3.21}$$

Hence

$$U_s = (G/\rho)^{\frac{1}{2}}; \quad \alpha = 0 \tag{5.3.22}$$

Typically U is of the order of 3000 m s^{-1}.

5.4 Propagation of Shear Waves in a Relaxing Medium: The Maxwell Model

We now evaluate the dependence of the viscoelastic properties of a medium on the frequency of the applied oscillatory shear strain in terms of the Maxwell model of viscoelasticity. Maxwell (1867) noted that a viscoelastic body could behave either as a Hookean solid or as a Newtonian liquid depending on the duration of a steady shear strain or the period of an oscillatory shear strain. He then incorporated Hooke's law (equation 5.1.1) and Newton's law (equation 5.1.2) into the same equation by proposing that the rate of change of the strain in a viscoelastic medium was given by

$$\frac{\partial \gamma}{\partial t} = \frac{1}{G_\infty} \frac{\partial \sigma}{\partial t} + \frac{\sigma}{\tau_s G_\infty} \tag{5.4.1}$$

Here G_∞ is the instantaneous or high-frequency rigidity modulus of the medium and τ_s is a relaxation time.

The meaning of equation (5.4.1) may be seen by considering the application at time $t = 0$ of a steady shear stress σ_0 to a Maxwell body. If the application of σ_0 occurs instantaneously, that is in a time much less than the relaxation time, $\partial \sigma / \partial t$ is very large, and equation (5.4.1) reduces to

$$\frac{\partial \gamma}{\partial t} = \frac{1}{G_\infty} \frac{\partial \sigma}{\partial t} \tag{5.4.2}$$

Integration of this equation gives

$$\gamma_\bullet' = \sigma_0 / G_\infty \tag{5.4.3}$$

Hence Hooke's law applies when $t \ll \tau$, and the body behaves as an elastic solid: the instantaneous stress produces an instantaneous elastic strain. At longer times viscous flow occurs. This may be seen by integrating (5.4.1) and inserting the boundary condition (5.4.3) for $t = 0$:

$$\gamma = \frac{\sigma_0}{G_\infty} + \frac{\sigma_0 t}{\tau_s G_\infty} \tag{5.4.4}$$

When $t \gg \tau$, the elastic contribution is negligible, and (5.4.4) becomes

$$\gamma = \sigma_0 t / \tau_s \, G_\infty \tag{5.4.5}$$

A strain linearly proportional to time as in (5.4.5) indicates viscous flow. Comparing (5.4.5) with Newton's equation (5.1.2), we see that the viscosity of a Maxwell body is the product of a relaxation time and an elastic modulus:

$$\eta_s = \tau_s \, G_\infty \tag{5.4.6}$$

The Maxwell equation of viscoelasticity (5.4.1) thus reproduces the behaviour of a viscoelastic body when subjected to a steady shear stress.

Consider now the application of a sinusoidal shear strain

$$\gamma = \gamma_0 \exp(i\omega t) \tag{5.4.7}$$

which gives rise to a sinusoidal shear stress. The time derivatives in equation (5.4.1) can be replaced by $i\omega$ as discussed in Section 2.4 to give

$$\sigma + i\omega\tau_s \, \sigma = i\omega\tau_s \, G_\infty \gamma \tag{5.4.8}$$

Hence the complex rigidity modulus G^* is

$$G^* = \frac{\sigma}{\gamma} = \frac{i\omega\tau_s \, G_\infty}{1 + i\omega\tau_s} \tag{5.4.9}$$

while the components of G^* are obtained by comparing real and imaginary parts of (5.4.9):

$$G' = \frac{G_\infty \omega^2 \tau_s^2}{1 + \omega^2 \tau_s^2}; \quad G'' = \frac{G_\infty \omega\tau_s}{1 + \omega^2 \tau_s^2} \tag{5.4.10}$$

In a Newtonian liquid ($\omega\tau_s \ll 1$),

$$G'' \approx G_\infty \omega\tau_s = \omega\eta_s \tag{5.3.18}$$

Similar expressions may be deduced for the components of η^*, J^* and Z_L from equations (5.3.9), (5.3.12) and (5.3.17):

$$\eta' = \frac{G_\infty \tau_s}{1 + \omega^2 \tau_s^2}; \quad \eta'' = \frac{G_\infty \omega\tau_s^2}{1 + \omega^2 \tau_s^2} \tag{5.4.11}$$

$$J' = 1/G_\infty = J_\infty; \quad J'' = 1/\omega\eta_s \tag{5.4.12}$$

$$\left.\begin{array}{l} R_L = (\rho G_\infty)^{\frac{1}{2}} \left\{ \dfrac{\omega^2 \tau_s^2 + \omega\tau_s(1 + \omega^2 \tau_s^2)^{\frac{1}{2}}}{2(1 + \omega^2 \tau_s^2)} \right\}^{\frac{1}{2}} \\[4mm] X_L = (\rho G_\infty)^{\frac{1}{2}} \left\{ \dfrac{-\omega^2 \tau_s^2 + \omega\tau_s(1 + \omega^2 \tau_s^2)^{\frac{1}{2}}}{2(1 + \omega^2 \tau_s^2)} \right\}^{\frac{1}{2}} \end{array}\right\} \tag{5.4.13}$$

The variation of G^*, η^* and Z_L with frequency is shown in Figure 5.4.1.

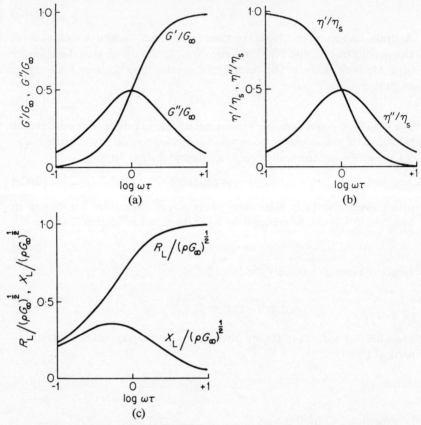

FIGURE 5.4.1. Variation with frequency of the normalized components of (a) the rigidity modulus, (b) the dynamic viscosity, (c) the shear mechanical impedance for a single Maxwell relaxation process.

5.5 Distributions of Relaxation and Retardation Times

The Maxwell model of viscoelasticity outlined in Section 5.4 assumes that the dependence upon frequency of the viscoelastic properties of a medium may be described in terms of a single relaxation time. Most substances have been found to require more than one relaxation time. In polymers, for example, a large number of discrete processes exist each having its individual relaxation time. If each of N processes contributes additively to G^*, we may generalize equation (5.4.9) to give

$$G^* = \sum_{j=1}^{N} G_j \frac{i\omega(\tau_\mathrm{s})_j}{1+i\omega(\tau_\mathrm{s})_j} \qquad (5.5.1)$$

G_j is the contribution of the jth process to G_∞ of the medium, and $(\tau_s)_j$ is the relaxation time of the jth process.

It is often convenient to describe viscoelastic relaxation in terms of a continuous distribution of relaxation times. Equation (5.4.9) is then written

$$G^* = G_\infty \int_0^\infty g(\tau_s) \frac{i\omega\tau_s}{1+i\omega\tau_s} d\tau_s \qquad (5.5.2)$$

$g(\tau_s)$ represents the distribution of relaxation times. $G_\infty g(\tau_s) d\tau_s$ is the contribution to G_∞ of the medium from those processes which have individual relaxation times in the range τ_s to $(\tau_s + d\tau_s)$. $g(\tau_s)$ is normalized such that

$$\int_0^\infty g(\tau_s) d\tau_s = 1 \qquad (5.5.3)$$

For a Maxwell body, $g(\tau_s)$ is a delta function, and (5.5.2) becomes identical to (5.4.9).

Separation of (5.5.2) into real and imaginary parts gives

$$\left.\begin{aligned}
G' &= G_\infty \int_0^\infty g(\tau_s) \frac{\omega^2 \tau_s^2}{1+\omega^2 \tau_s^2} d\tau_s \\
G'' &= G_\infty \int_0^\infty g(\tau_s) \frac{\omega\tau_s}{1+\omega^2 \tau_s^2} d\tau_s
\end{aligned}\right\} \qquad (5.5.4)$$

The static viscosity η_s of the medium is the value of G''/ω when $\omega\tau_s \ll 1$:

$$\eta_s = G''/\omega = G_\infty \int_0^\infty \tau_s g(\tau_s) d\tau_s = G_\infty \bar{\tau}_s \qquad (5.5.5)$$

where $\bar{\tau}_s$ is an average relaxation time (cf. 5.4.6).

Because the observed range of relaxation times often covers several decades, the distribution of relaxation times is sometimes expressed on a logarithmic scale (Ferry, 1961). Exact mathematical relationships are available for calculating the distribution function $g(\tau_s)$ from experimental values of G' or G'' (Gross, 1968). In practice, approximation methods are usually more convenient (Alfrey, 1948; Schwarzl and Struik, 1968).

J^* of a liquid is rarely described by the simple Maxwell relation of equation (5.4.12). The more general expression is (Gross, 1968)

$$J^* = J' - iJ'' = J_\infty - \frac{i}{\omega\eta_s} + (J_0 - J_\infty) \int_0^\infty f(\tau_s) \frac{i\omega}{1+i\omega\tau_s} d\tau_s \qquad (5.5.6)$$

whence

$$J' = J_\infty + (J_0 - J_\infty) \int_0^\infty f(\tau_s) \frac{1}{1 + \omega^2 \tau_s^2} d\tau_s$$

$$J'' = \frac{1}{\omega \eta_s} + (J_0 - J_\infty) \int_0^\infty f(\tau_s) \frac{\omega \tau_s}{1 + \omega^2 \tau_s^2} d\tau_s$$

$$(5.5.7)$$

Here J_0 is the long time, low-frequency compliance and $f(\tau_s)$ is the distribution of viscoelastic retardation times. The relations between $f(\tau_s)$ and the distribution of relaxation times $g(\tau_s)$ are discussed by Gross (1968) and Schwarzl and Struik (1968). Mathematically $f(\tau_s)$ not $g(\tau_s)$ has the same form as the distribution of dielectric relaxation times (McCrum, Read and Williams, 1967). Plots of J' and J'' are shown in Figure 5.5.1. At long times (low frequencies)

$$J' = J_0; J'' = 1/\omega \eta_s \qquad (5.5.8)$$

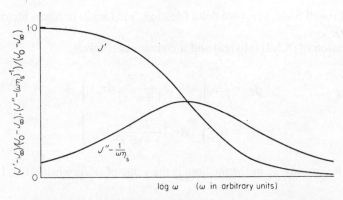

FIGURE 5.5.1. Variation with frequency of the normalized components of the shear compliance for a liquid having a distribution of retardation times.

Many empirical expressions have been proposed for the distribution of relaxation times $g(\tau_s)$. The Gaussian distribution (Yager, 1936) is

$$g(\tau_s) d\tau_s = (b/\pi)^{\frac{1}{2}} \exp(-b^2 z^2) dz \qquad (5.5.8)$$

where

$$z = \log(\tau_s/\tau_s') \qquad (5.5.9)$$

This distribution has a maximum at τ_s' and is symmetrical when plotted against the logarithm of frequency (Figure 5.5.2). The parameter b determines the width of the distribution while τ_s' determines the position of the maximum. The components of G^* can be calculated by substituting (5.5.8) into (5.5.4), although the integrals cannot be evaluated analytically.

Another useful distribution is the asymmetric Davidson–Cole (1951) function which was originally developed for dielectric relaxation. In viscoelastic work it has been customary to use this distribution to describe G^* by rewriting equation (5.4.9) as

$$G^* = \frac{i\omega\tau_s\, G_\infty}{(1+i\omega\tau_s)^\beta}$$
(5.5.10)

β determines the width of the distribution, $\beta = 1$ corresponding to a single relaxation time. $g(\tau_s)$, G' and G'' can be evaluated analytically from

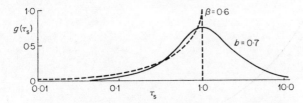

FIGURE 5.5.2. Distribution of relaxation times $g(\tau_s)$ for a Davidson–Cole distribution with $\beta = 0.6$ (---) and a Gaussian distribution with $b = 0.7$ (—).

(5.5.10), and are quoted by Litovitz and Davis (1965). This distribution is asymmetric with a sharp cut-off at long times (Figure 5.5.2).

Other distributions which have been applied in dielectric studies are those of Cole and Cole (1941), Fuoss and Kirkwood (1941) and Williams and Watts (1970). It must be emphasized that all such distribution functions are purely empirical: they form a convenient way of describing experimental results mathematically, but are in no way a theoretical treatment of the relaxation process. This means that such functions could be used to describe either $g(\tau_s)$ or $f(\tau_s)$. Since $f(\tau_s)$ resembles mathematically the distribution of dielectric relaxation times more closely than does $g(\tau_s)$, it would seem reasonable to apply the empirical dielectric functions to the distribution of viscoelastic retardation times. In practice, discussions of the viscoelastic behaviour of liquids under alternating strain are usually confined to $g(\tau_s)$ because of the difficulty of determining J_0 experimentally.

5.6 The Stress Relaxation Function and the Creep Compliance

The previous sections have referred only to oscillatory shear. Equivalent information about the viscoelastic properties of a medium may be obtained from the step-function experiments of stress relaxation and creep. In stress relaxation the sample is suddenly strained to a given deformation γ_0 and the stress σ required to maintain this strain is measured as a function of time. The expression for stress relaxation is

$$\sigma = [G_0 + \psi(t)]\gamma \quad \begin{cases} \gamma = 0, & t < 0 \\ \gamma = \gamma_0, & t \geq 0 \end{cases} \tag{5.6.1}$$

G_0 is the static rigidity modulus of the medium which equals zero in many cases. $\psi(t)$ is the stress relaxation function, and its behaviour is illustrated in Figure 5.6.1.

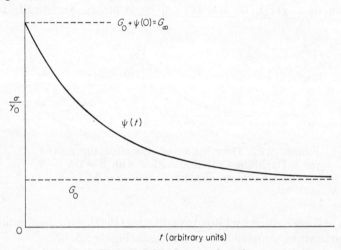

FIGURE 5.6.1. Stress relaxation.

In creep a stress σ_0 is suddenly applied and maintained constant while the resulting strain is measured as a function of time:

$$\gamma = [J_\infty + (t/\eta_s) + J_r(t)]\sigma \quad \begin{cases} \sigma = 0, & t < 0 \\ \sigma = \sigma_0, & t \geq 0 \end{cases} \tag{5.6.2}$$

$$= J(t)\sigma$$

$J_\infty (= 1/G_\infty)$ is the instantaneous elastic compliance, $J_r(t)$ is the recoverable compliance, t/η_s is the contribution to the strain arising from viscous flow, and $J(t)$ is the creep compliance. The time dependence of equation (5.6.2) is shown in Figure 5.6.2.

To calculate $\psi(t)$ for a Maxwell body, we assume that γ_0 is applied instantaneously at $t = 0$. For times $t > 0$, $\partial\gamma/\partial t$ is zero, and equation (5.4.1) reduces to

$$0 = \frac{1}{G_\infty}\frac{\partial\sigma}{\partial t} + \frac{\sigma}{\tau_s G_\infty} \tag{5.6.3}$$

Integrating this equation, and assuming that the instantaneous stress on applying γ_0 is σ_0 where

$$\sigma_0 = [G_0 + \psi(0)]\gamma_0 = G_\infty \gamma_0 \qquad (5.6.4)$$

we obtain

$$\sigma = \sigma_0 \exp(-t/\tau_s) \qquad (5.6.5)$$

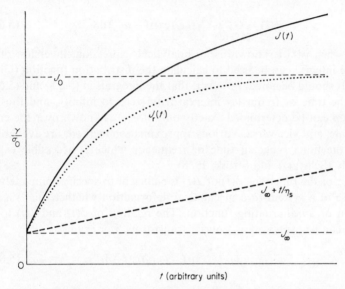

FIGURE 5.6.2. Creep in a liquid having a distribution of retardation times.

Here the stress relaxes exponentially to zero. Since $G_0 = 0$ for a Maxwell body, (5.6.1) yields

$$\psi(t) = (\sigma_0/\gamma_0)\exp(-t/\tau_s) = G_\infty \exp(-t/\tau_s) \qquad (5.6.6)$$

that is, the stress relaxation function is an exponential.

An expression for $J(t)$ for a Maxwell body may be obtained from equations (5.4.1) and (5.6.2) by a similar process. It transpires that $J_r(t) = 0$, and hence

$$J(t) = J_\infty + t/\eta_s \qquad (5.6.7)$$

In the general case where a distribution of relaxation times $g(\tau_s)$ exists, equation (5.6.6) becomes

$$\psi(t) = G_\infty \int_0^\infty g(\tau_s)\exp(-t/\tau_s)\,d\tau_s \qquad (5.6.8)$$

A similar expression may be obtained for $J(t)$ in terms of the distribution of retardation times. Equation (5.6.8) may conveniently be represented as a Laplace integral by introducing a relaxation frequency $\omega_s = 1/\tau_s$ and a frequency function $N(\omega_s)\,d\omega_s$ where

$$N(\omega_s) = g(1/\omega_s)/\omega_s^2 \qquad (5.6.9)$$

Equation (5.6.8) is then

$$\psi(t) = G_\infty \int_0^\infty N(\omega_s)\exp(-t\omega_s)\,d\omega_s \qquad (5.6.10)$$

Hence when $\psi(t)$ is known as an analytical expression, inversion of the Laplace integral can be used to calculate the distribution function (Gross, 1968). It should be noted, however, that the integrals in (5.6.8) and (5.6.10) cover the time or frequency interval from zero to infinity, and thus the spectrum can be determined exactly only if $\psi(t)$ is known over the entire time scale, and vice versa. Various approximation methods are available to utilize the finite range of time or frequency which is accessible experimentally (Schwarzl and Struik, 1968).

Either of the functions $\psi(t)$ or $J(t)$ is sufficient to specify completely the response of a system to a mechanical deformation whether this is a step function or an alternating function. The relation of $\psi(t)$ and $J(t)$ to G^* and J^* can be expressed as Fourier transforms. For example

$$G^* = i\omega \int_0^\infty \exp(-i\omega t_s)\psi(t_s)\,dt_s \qquad (5.6.11)$$

or

$$\left. \begin{aligned} G' &= \omega \int_0^\infty \psi(t_s)\sin(\omega t_s)\,dt_s \\ G'' &= \omega \int_0^\infty \psi(t_s)\cos(\omega t_s)\,dt_s \end{aligned} \right\} \qquad (5.6.12)$$

The integrals in (5.6.12) allow the calculation of the components of G^* from the relaxation function $\psi(t)$. Fourier inversion of these integrals allows calculation of the relaxation function from G^*:

$$\left. \begin{aligned} \psi(t) &= \frac{2}{\pi}\int_0^\infty \frac{G'}{\omega}\sin\omega t\,d\omega \\ \psi(t) &= \frac{2}{\pi}\int_0^\infty \frac{G''}{\omega}\cos\omega t\,d\omega \end{aligned} \right\} \qquad (5.6.13)$$

Equations (5.6.12) and (5.6.13) relate the behaviour under static conditions to that during a dynamic experiment.

In this chapter the response of a viscoelastic medium to a shear deformation has been discussed in terms of three types of function: the complex moduli G^* and J^*, the complex viscosity η^* and the complex shear mechanical impedance Z_L; the distribution of relaxation times $g(\tau_s)$ and the distribution of retardation times $f(\tau_s)$; and the stress relaxation function $\psi(t)$ and the creep compliance $J(t)$. The interrelations between these functions are shown diagrammatically in Figure 5.6.3: a complete list of the interrelations is given by Schwarzl and Struik (1968).

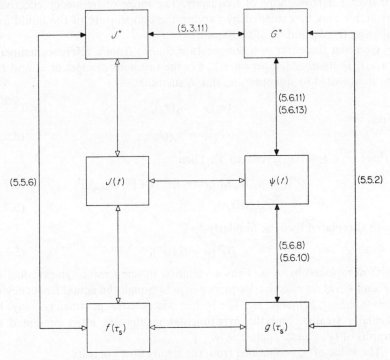

FIGURE 5.6.3. Structure of the theory of linear viscoelasticity: full arrows indicate interrelations discussed in text; open arrows indicate interrelations not discussed; numbers indicate equations describing an interrelation.

Any one of these functions can in principle be calculated from any other if accurate experimental observations are available over the entire range of time or frequency from zero to infinity. In practice, an experiment covers a finite and often very limited range of time or frequency with modest accuracy: for example, experiments with high frequency shear waves usually study the short-time behaviour of a medium while stress relaxation and creep investigate longer times. Hence a combination of different experimental techniques is necessary to define the viscoelastic behaviour of a medium completely, and studies with only one technique over a limited range have often led to erroneous conclusions.

5.7 The Method of Reduced Variables: Time–Temperature Superposition

Many of the experimental techniques for studying the viscoelastic behaviour of liquids under alternating strain operate at a single frequency or over a narrow range of frequency. The range of frequency effectively available may be extended by varying the temperature of the liquid and altering G_∞, τ_s and η_s of the liquid.

Consider the effect of a temperature change from a reference temperature T_r to another temperature T. Let the resultant changes in τ_s and G_∞ be represented by quantities α_r and β_r such that

$$(\tau_s)_T = \alpha_r(\tau_s)_r \tag{5.7.1}$$

and

$$(G_\infty)_T = \beta_r(G_\infty)_r \tag{5.7.2}$$

where $(\tau_s)_r$ and $(G_\infty)_r$ refer to T_r. Then

$$(G')_T = \beta_r(G_\infty)_r[\alpha_r\,\omega(\tau_s)_r]^2/\{1 + [\alpha_r\,\omega(\tau_s)_r]^2\} \tag{5.7.3}$$

$$= \beta_r(G')_r \tag{5.7.4}$$

with ω replaced by $\alpha_r\,\omega$. Similarly

$$(G'')_T = \beta_r(G'')_r \tag{5.7 5}$$

with ω replaced by $\alpha_r\,\omega$. Thus a variation in temperature gives values of G' and G'' at an effective frequency $\alpha_r\,\omega$ although the actual frequency of measurement remains ω. The other viscoelastic parameters may be similarly treated, and the experimental results are often presented as graphs of G', η', etc. against $\alpha_r\,\omega$.

The value of α_r is obtained from the steady flow viscosity:

$$(\eta_s)_T = (\tau_s)_T(G_\infty)_T \quad \text{by (5.4.6)}$$

$$= \alpha_r\beta_r(\tau_s)_r(G_\infty)_r$$

$$= \alpha_r\beta_r(\eta_s)_r \tag{5.7.6}$$

The ratio of the steady flow viscosity at two temperatures gives the product $\alpha_r\beta_r$. β_r can usually be found with sufficient accuracy by extrapolating the observed G_∞ values to higher temperatures (Section 9.2). A particularly convenient way of presenting the experimental results is to plot G', η', etc. against $\log(\omega\eta_s/G_\infty)$.

Equation (5.7.6) is based on the Maxwell relation (5.4.6). The method of reduced variables (or time–temperature superposition) may also be applied where a distribution of relaxation times exists provided that all the relaxation times in the distribution have the same dependence on temperature. Although this condition is probably not strictly true, it appears to hold approximately, and this method has been widely used to extend the frequency range of a given technique (Ferry, 1961). The method has also been applied to variations of pressure (Philippoff, 1963). The application of this method to the particular case of polymers is discussed in Section 7.3.

CHAPTER 6

EXPERIMENTAL TECHNIQUES FOR STUDYING THE VISCOELASTIC PROPERTIES OF LIQUIDS

6.1 Introduction

In this Chapter we shall discuss the oscillatory methods which are available for studying the viscoelastic properties of liquids and solutions. We shall consider a liquid to be a substance which may be poured into the measuring apparatus, and we shall not discuss techniques for studying the behaviour of glasses, rubbers or gels. We shall also omit apparatus for studying non-linear effects.

Few techniques cover more than two decades of frequency, and since viscoelastic relaxation in liquids above their glass transition temperatures occurs with relaxation times ranging from several minutes to 10^{-12} s, a large number of different experimental arrangements will be required. In simple liquids in which the relaxation region extends over only one decade of frequency, a single technique may be adequate to define the relaxation region completely. In polymer liquids, however, viscoelastic relaxation occurs over many decades of frequency, and no single technique can cover more than a small portion of this.

In principle the most direct way of determining the viscoelastic properties of a liquid is to apply a shear stress and to follow the resultant strain as a function of time. This method is only applicable to very viscous liquids, however. The wave propagation methods commonly used with ultrasonic longitudinal waves (Chapter 3) are also of limited application to shear waves, since the latter are rarely propagated more than a few wavelengths into the liquid. Shear waves do have a loading effect upon the generating transducer, however, and their properties may be deduced from the changes in the electrical impedance of the transducer. Various modifications of this method are widely used at frequencies below 1 MHz. Above this frequency, the small size of the transducers necessitates their attachment to a solid delay line (Section 3.7) through which the shear waves are propagated: the properties of the liquid are then obtained by studying the reflexion of the shear waves at the delay line–liquid interface.

The following sections give an account of the more successful methods of studying the viscoelastic properties of liquids in the frequency range 10^{-4}–10^{10} Hz. Often it is only possible to study a given liquid over a limited frequency range. It is then customary to extend the effective frequency range by the time–temperature superposition principle in which a change in the temperature and hence in the viscosity of the liquid sample is assumed to be equivalent to a change in the frequency of the applied shear waves (Sections 5.7 and 7.3).

6.2 Propagation of Shear Waves in Liquids: 4–5000 Hz

It was shown in Section 5.3 that shear waves are heavily damped in liquids and that they do not propagate more than a few wavelengths from the source. At high frequencies this propagation distance is usually too small for measurements to be made of the velocity and attenuation of shear waves by the techniques commonly used to study ultrasonic wave propagation (Chapter 3). At frequencies below 5 kHz, however, propagation techniques have been used to study the viscoelastic behaviour of polymer liquids (Ferry, 1941; Adler, Sawyer and Ferry, 1949).

The liquid under investigation is contained in a rectangular cell, and shear waves are generated by a loudspeaker which causes a thin plate to vibrate vertically in its own plane. If the liquid becomes birefringent on straining, the shear waves in the liquid can be detected by a suitable optical system: for example, a stationary pattern will be observed when the cell and a double quartz wedge are placed between crossed polaroids and illuminated with a stroboscope synchronized with the vibration frequency of the plate. The wavelength and velocity U_s of the shear waves is then obtained directly, while the absorption coefficient α is calculated from the displacements in the optical pattern. The components of the rigidity modulus G^* are then found from equations (5.3.6) and (5.3.7).

The upper frequency limit of this technique is governed by the large attenuation at small wavelengths. The lower frequency limit occurs when the wavelength becomes comparable to the size of the liquid container and standing waves cause unreliable results to be obtained. The method has been applied to the study of a variety of concentrated polymer solutions (Ashworth and Ferry, 1949). It has the advantage of simplicity, and unlike most other low frequency techniques it is independent of the size and shape of the containing vessel. The main disadvantage is the rather poor accuracy with which the viscoelastic functions may be determined: for the dynamic viscosity, for instance, the experimental error is typically of the order of 15%.

6.3 Direct Measurement of Stress and Strain: 10^{-4}–10^2 Hz

Another simple method of studying the viscoelastic properties of liquids at very low frequencies is to apply a sinusoidal stress to the liquid and follow the resulting strain as a function of time. From equation (5.3.8) the ratio of peak stress to peak strain is given by

$$| G^* | = (G'^2 + G''^2)^{\frac{1}{2}} \tag{6.3.1}$$

and the phase angle δ is given by

$$\tan \delta = G''/G' \tag{6.3.2}$$

Hence G' and G'' may be calculated.

One of the most direct systems for making measurements of this type is that of Morrisson, Zapas and DeWitt (1955) (Figure 6.3.1). The liquid to

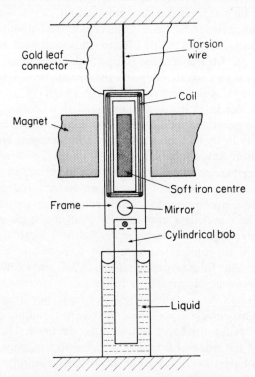

FIGURE 6.3.1. Forced oscillation apparatus of
Morrisson, Zapas and DeWitt (1955).

be investigated is contained in a coaxial cylinder with only a small gap between the cylinders. The inner cylinder is driven in forced oscillation by a coil carrying an alternating current of the desired frequency suspended in a d.c. magnetic field, and the whole system is suspended by a torsion wire. The amplitude of the stress is determined by a light beam reflected from the mirror of a galvanometer in series with the coil, while the strain is observed by light from a mirror mounted on the coil. The system must first be calibrated to determine the moment of inertia of the inner cylinder, the stiffness of the torsion wire and the torque produced by unit current in the coil.

This method becomes inaccurate when tan δ is small. It is then preferable to determine the decrement of free vibrations of the inner cylinder.

This technique is very successful for studying the dynamic viscosity of polymer solutions in the viscosity range 10^{-1}–10^5 N s m^{-2}. Determinations of the viscosity of Newtonian liquids may be made to within a few per cent of the true value, while rigidities as low as 10^{-1} N m^{-2} can be measured.

6.4 Transducer Measurements of Stress and Strain: 10^{-2}–1.5×10^3 Hz

At frequencies above 100 Hz, the direct measurement of stress and strain becomes impracticable, since the various mechanical components of the system cannot be made sufficiently stiff to prevent their breaking into vibration. The preferred method of studying the viscoelastic behaviour of liquids is then based on an electromagnetic driving and detecting system in which only the surfaces in contact with the liquid sample need be rigid. A number of different experimental systems have been devised, each with its own particular advantages.

In the apparatus of Birnboim and Ferry (1961) the liquid under investigation is contained in a rigid outer cylinder. To a solid inner cylinder is attached a coil carrying an alternating current of the desired frequency and suspended in a permanent magnetic field. This causes the inner cylinder to oscillate along its axis with very low amplitude, and subjects the liquid to an oscillating shear stress. The viscoelastic properties of the liquid may then be determined from measurements of the electrical impedance of the coil. This axial oscillation (annular pumping motion) may be contrasted with the oscillatory rotation of the system shown in Figure 6.3.1.

If the current flowing in the coil is i, the magnetic flux density is B and the length of wire in the coil is l, then the force F exerted on the moving system is

$$F = Bli \tag{6.4.1}$$

The inner cylinder will respond by moving with velocity v, where v is given by the definition of mechanical impedance:

$$v = Bli/Z_M \tag{6.4.2}$$

The motion of the coil generates a back e.m.f. given by

$$e_B = Blv = B^2 l^2 i/Z_M \tag{6.4.3}$$

This back e.m.f. opposes the applied e.m.f. $Z_0 i$, where Z_0 is the electrical impedance of the coil at rest. If the resultant e.m.f. is Zi, where Z is the electrical impedance of the coil while in motion, then

$$Zi = Z_0 i - e_B \qquad (6.4.4)$$

or

$$Z - Z_0 = B^2 l^2 / Z_M \qquad (6.4.5)$$

The impedances Z and Z_0 may be measured by placing the driving coil in one arm of an impedance bridge. To determine Z_0, a yoke is used to clamp the coil rigidly against its support. The apparatus constant $B^2 l^2$ is obtained by measuring the impedance of the system over a range of frequencies with no liquid present. Hence the mechanical impedance of the moving system Z_M can be determined.

Z_M is related to the components of the shear rigidity modulus G^* of the liquid by the expression

$$Z_M = a \frac{G''}{\omega} + i \left[\omega(M + \rho\beta a) - \frac{S_M}{\omega} - \frac{aG'}{\omega} \right] \qquad (6.4.6)$$

Here M is the mass of the moving system, S_M is the ratio of force to displacement in the springs on which the inner cylinder is mounted, and a and β are geometrical constants for the system. Once the apparatus has been calibrated, the viscoelastic properties of the liquid may be determined from measurements of the electrical impedance of the coil in motion Z by equations (6.4.5) and (6.4.6).

This impedance bridge method has been used in the frequency range 2·5–400 Hz. At low frequencies (0·01–5 Hz) this method of measurement is unreliable and it is more convenient to determine the ratio of the maximum amplitudes of the applied stress and resultant strain, together with the phase angle (cf. Section 6.3). The stress is obtained from the current flowing in the coil, and the strain by following the motion of the inner cylinder with a differential transformer. To determine the phase difference between these two signals, they are clipped at a low level and the resulting square waves passed through a differential amplifier. The phase is then obtained from the time separation of the resulting rectangular pulses.

The Birnboim and Ferry (1961) apparatus can be operated over a continuously variable range of frequencies from 0·01 to 400 Hz (Tschoegl and Ferry, 1963). Only 1 cm^3 of liquid is required and the apparatus is capable of high precision: the viscosity of a Newtonian oil, for example, can be determined to within 0·2% of its true value. The apparatus has been extensively used to study the viscoelastic properties of polymer solutions, but is limited to solutions where the viscosity exceeds 0·1 N s m^{-2}. The sensitivity is considerably increased if the solid moving rod is replaced by a thin-walled tube: the liquid is then sheared both within and without the tube (Simmons, 1966).

Above 400 Hz the apparatus of Birnboim and Ferry (1961) becomes unreliable because of mechanical resonance. Lamb and Lindon (1967) have developed an electromagnetic, torsionally vibrating system which covers the frequency range 20–1500 Hz. The liquid under investigation is sheared in the annulus between the inner surface of a torsionally oscillating cylindrical cup and a fixed cylinder of smaller radius. A four-pole d.c. electromagnet surrounds the torsionally vibrating cup, the latter being suspended by four steel strips symmetrically disposed around the outer surface. The driving and pick-up coils have the same radius and are fixed axially in the cup at right angles to each other. When an alternating current is passed through the driving coil, the cross-magnetic field of the electromagnet causes the cup to oscillate torsionally at the frequency of the driving current. Cylindrical shear waves travel inward from the cup and are reflected at the fixed cylinder so that standing waves are set up in the liquid annulus. The resulting motion of the cup is detected by the pick-up coils.

The shear mechanical impedance of the liquid may be calculated from the relative amplitudes and phase of the signals in the input and output coils. The detailed analysis is considerably complicated by the presence of standing torsional waves in the liquid. The lower limit of 20 Hz is set by the electrical measuring system, while the upper limit of 1500 Hz is determined by mechanical resonance of the cup which contains the liquid. The overall reproducibility of the components of the rigidity modulus of the liquid is $\pm 5\%$, which compares unfavourably with that of the Birnboim and Ferry (1961) instrument. The Lamb and Lindon (1967) apparatus has the advantage, however, that it generates a pure torsional wave in the liquid: this removes the uncertainty caused by compressional wave generation in the annular pumping apparatus. The instrument has been used to study the viscoelastic properties of liquids in which the steady-flow viscosity exceeds $0 \cdot 1$ N s m^{-2}.

6.5 Impedance Methods for Low Viscosity Liquids: 10^2–10^4 Hz

The non-resonant systems of the previous section can be driven at any desired frequency within their operating range. As a result, however, they are insufficiently sensitive to study low viscosity liquids. The two most suitable techniques for this application are the torsion pendulum of Sittel, Rouse and Bailey (1954) and the crystal tuning fork of Mason (1958).

The torsion pendulum (Figure 6.5.1) consists of a hollow lower disc and a massive upper disc connected by a torsion rod, the dimensions of which are chosen to give the desired resonant frequency. The liquid to be investigated is contained in the lower disc which can be excited into torsional oscillation by the action of an external alternating magnetic field upon a magnet fixed to the lower disc. The inertia of the liquid allows the tangential motion of the disc surface to generate a shear wave in the liquid. This is effectively a plane wave since it is damped out in a distance which is very small compared to the radius of curvature of the disc surface.

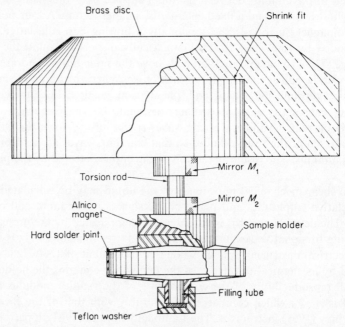

FIGURE 6.5.1. Hollow torsion pendulum of Sittel, Rouse and Bailey (1954).

The motion of the disc is studied by reflecting a light beam onto a photomultiplier from a mirror attached to the lower disc, the light reflected from the upper disc serving as a reference. To make a measurement, the oscillation of the pendulum is allowed to die away freely and the resonant frequency ω and the logarithmic decrement of the motion α determined. This is repeated with the liquid under investigation, and the components of the shear mechanical impedance of the liquid are obtained from the changes in ω and α.

The oscillations of the pendulum obey the equation

$$I\ddot{\theta} + Z_t\dot{\theta} + k_r\theta = 0 \qquad (6.5.1)$$

where I is the moment of inertia of the lower disc, θ is its instantaneous angular displacement, and k_r is the torque per unit deflexion of the torsion rod. Z_t is the total impedance of the moving system and is given by

$$Z_t = R_i + (R_m + iX_m) \tag{6.5.2}$$

Here R_i is the resistance arising from internal friction in the metal, and $(R_m + iX_m)$ is the moment of impedance of the liquid: this is given by the integral over the internal surface of the hollow disc of the product $Z_L r^2 \, dA$, where Z_L is the shear mechanical impedance of the liquid, and r is the distance of the area dA from the axis of rotation.

The solution to equation (6.5.1) is

$$\theta = \theta_0 \exp(i\omega t) \exp(-\alpha t) \tag{6.5.3}$$

When no liquid is contained in the disc, $Z_t = R_i$, and

$$\alpha = \alpha_0 = R_i/2I, \tag{6.5.4}$$

and

$$\omega^2 = \omega_0^2 = (k_r/I) - \alpha_0^2 \approx k_r/I \tag{6.5.5}$$

These values of R_i and k_r/I can then be used when the disc is full of liquid. Under these conditions α and $\Delta\omega$ ($= \omega_0 - \omega$) are usually small compared with ω, and

$$\alpha = (R_i + R_m)/2I \tag{6.5.6}$$

and

$$\Delta\omega = X_m/2I \tag{6.5.7}$$

Hence the components of Z_m may be obtained from the change in frequency and rate of decay of oscillation of the disc, and from Z_m the shear mechanical impedance of the liquid Z_L is obtained.

This system has been successfully used to study the viscoelastic properties of dilute polymer solutions of viscosity of the order of 10^{-3} N s m^{-2} in the frequency range 200–2500 Hz.

An alternative method for low viscosity liquids which is capable of high precision is the quartz crystal tuning fork of Mason (1958). This is a thin, flat piece of crystal quartz cut in the shape of a tuning fork and silvered on the opposite flat faces. When an alternating voltage is applied to these faces at a frequency corresponding to a resonant frequency of the system, the thin, wide arms oscillate in their own planes and set up plane shear waves in the liquid. The components of the shear mechanical impedance of the liquid may be calculated from the change in the resonant frequency and in the resistance at resonance in a manner similar to that for the torsional crystal discussed in Section 6.6. The crystal tuning fork gives very precise measurements of the shear impedance of Newtonian liquids in the frequency range 5×10^2–10^4 Hz and it has also been used in the study of polymer solutions. Since the only connexions to the tuning fork are electrical, it should be capable of working over wide ranges of temperature and pressure.

6.6   The Torsional Quartz Crystal: 10–120 kHz

One of the most successful techniques for studying the viscoelastic behaviour of liquids is the resonant torsional quartz crystal, originally devised by Mason (1947). It consists of a cylinder of crystal quartz cut with the X axis along the length of the cylinder (Figure 6.6.1). The crystal

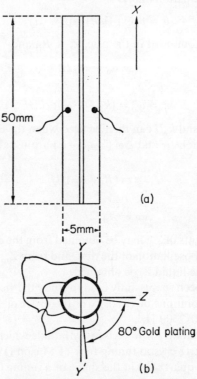

FIGURE 6.6.1. Longitudinal (a) and transverse (b) sections through a torsionally vibrating quartz crystal resonant at 38·6 kHz.

is gold plated in four strips along its length, and opposite electrodes are connected. When an alternating signal is applied to the electrodes, the electric fields in the Y direction on either side of the line YY' are in opposite directions and produce two opposing shear stresses. These induce

a torsional oscillation of the cylinder. If the frequency of the applied signal
is adjusted to coincide with a natural resonance frequency of the crystal,
then oscillations of appreciable amplitude are obtained. When the crystal
is immersed in a liquid, a cylindrical shear wave is propagated into the
liquid.

The viscoelastic properties of the liquid may be deduced from the
change in the electrical impedance of the crystal on immersion in the
liquid. Initially the resonant frequency and the resistance at resonance of
the crystal are determined *in vacuo*, or, in practice, in air. The crystal is
then immersed in the liquid to be investigated, and the resonant frequency
and the resistance at resonance redetermined. Mason (1947) has shown by
considering the equivalent electrical circuit of the crystal that the changes
observed in these quantities are related to the shear mechanical impedance
of the liquid Z_L ($= R_L + iX_L$):

$$\Delta R = \frac{r}{2\pi f_R^2 C_0 I}\left[a^3 + \frac{a^4}{l}\right] R_L = K_1 R_L \qquad (6.6.1)$$

$$\Delta f = -\frac{1}{2I}\left[a^3 + \frac{a^4}{l}\right] X_L = -K_2 X_L \qquad (6.6.2)$$

ΔR and Δf are the changes in the resistance at resonance and the resonant
frequency of the crystal, r is the ratio of the capacitances of the equivalent
electrical circuit of the torsional crystal, f_R is the frequency of resonance,
C_0 is the static capacitance of the crystal, $I = \pi \rho_Q a^4/2$ is the moment of
inertia of the crystal, ρ_Q is the density of crystal quartz, a is the radius of
the crystal and l its length. The proportionality constants K_1 and K_2 may
be calculated for a given crystal. It is preferable, however, to obtain these
constants by calibration with a Newtonian liquid in which

$$R_L = X_L = (\pi f \eta_s \rho)^{\frac{1}{2}} \qquad (5.3.19)$$

While there is good agreement between the two values of K_1 so obtained,
the calculated values of K_2 are generally several per cent lower than the
calibrated values. This discrepancy is probably caused by the electrodes
and mounting wires of the crystal. The liquid impedance obtained in this
way is the impedance which it presents to cylindrical shear waves: a small
correction must be applied to convert this to the plane wave impedance
(equation 6.7.2).

Figure 6.6.2 is a diagram of the electrical measuring system used by
Barlow and coworkers (1961). The crystal forms one arm of a radio-
frequency bridge which is driven by a variable frequency high stability
oscillator. The output from the bridge is amplified and displayed on an
oscilloscope. The capacitance of the crystal is first measured at a frequency

far from resonance. This value is close to the capacitance of the crystal at resonance, and the resistance and frequency at resonance may then be obtained by successive adjustments until cancellation is achieved. This procedure is carried out first in air, and is then repeated with the liquid to be investigated.

The resonant torsional crystal technique has been used to study the viscosity and viscoelastic relaxation of liquids and polymers over wide

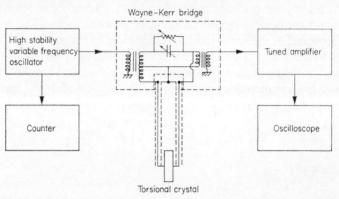

FIGURE 6.6.2. Measuring system using resonant torsional crystal
(after Barlow and coworkers, 1961).

ranges of temperature and pressure. In the Newtonian region values obtained for the viscosity of liquids agree with those of other methods to $\pm 1\%$, although with special precautions the error can be decreased to about $\pm 0.2\%$ (Diller, 1965). The components of the shear mechanical impedance of liquids are usually accurate to about $\pm 1\%$. There is no lower viscosity limit for this instrument, and it is sufficiently sensitive to measure the viscosity of a gas. The technique is only applicable to liquids with viscosities below 1 N s m^{-2}, however, since the resistance at resonance of higher viscosity liquids is too large to measure accurately. The frequency of operation of the crystal is confined to the region 10–120 kHz by the physical dimensions of the crystal: the large crystals required for lower frequencies are expensive, while at high frequencies it is difficult to grind the crystal accurately and yet maintain a sufficiently large length to diameter ratio to produce a pure torsional motion. Although the crystal can be driven at harmonics of its fundamental resonant frequency, some sensitivity is lost in the process.

6.7 Travelling Wave Technique: 10–200 kHz

This technique was originally devised by McSkimin (1952) as a modification of the resonant torsional crystal technique (Section 6.6) for use at higher viscosities. A torsional crystal is rigidly attached to one end of a solid rod as shown in Figure 6.7.1. When the crystal is excited by an alternating voltage, the torsional waves generated are transmitted along the rod. They experience a small attenuation and also a phase shift equal to $s\omega/U_t$, where s is the distance travelled along the rod, ω is the circular frequency, and U_t is the velocity of propagation of torsional waves in the rod. When the rod is surrounded by liquid, the loading on the surface

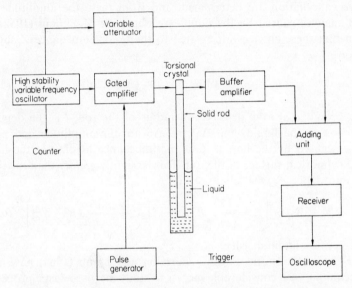

FIGURE 6.7.1. Measuring system for travelling torsional wave technique using piezoelectric excitation (after Barlow and coworkers, 1961).

increases the attenuation and alters the phase of the torsional waves. From the changes in these two quantities, the components of the liquid impedance may be derived.

It is convenient to use a single torsional crystal to transmit and receive the torsional waves in the rod, and this is best achieved by employing a pulse technique. Figure 6.7.1 is a block diagram of the electrical system used by Barlow and coworkers (1961) in their modification of the technique. The output from a high stability oscillator is gated to provide pulses of about 0·5 ms duration and 100 V amplitude which are applied to the torsional crystal. If the pulse length is shorter than the transit time of the torsional waves in the rod, a discrete pulse of torsional waves will be reflected up and down the rod many times. At each reflexion at the crystal, part of the elastic energy is converted into electrical energy and an exponentially decaying train of pulses is produced. This is added to an attenuated continuous wave signal from the oscillator, and the resultant amplified and displayed on the oscilloscope. By suitable adjustment of the attenuator setting and the oscillator frequency, the continuous wave and received signals may be arranged to be of the same amplitude but exactly 180° out of phase so that they cancel. On surrounding the rod by liquid, the changes in attenuator setting and oscillator frequency necessary to restore cancellation are determined, and from these the amplitude and phase changes caused by the liquid are obtained. McSkimin (1952) has shown that these changes are related to the liquid impedance Z_c by the relation

$$Z_c = R_c + iX_c = \frac{\rho_r U_t a}{2l}(\Delta\alpha + i\Delta\beta) \qquad (6.7.1)$$

where ρ_r and a are the density and radius of the rod, l is the depth of immersion in the liquid, and $\Delta\alpha$ and $\Delta\beta$ are the amplitude and phase changes caused by the liquid. Z_c is the impedance loading of the liquid on the cylindrical surface of the rod. This is slightly greater than the plane wave impedance Z_L, the relation being

$$Z_L = R_L + iX_L = \left[R_c - \frac{3}{\omega\rho_1 a}R_c X_c\right] + i\left[X_c + \frac{3}{2\omega\rho_1 a}(R_c^2 - X_c^2)\right] \qquad (6.7.2)$$

where ρ_1 is the liquid density.

McSkimin (1952) and Barlow, Harrison and Lamb (1964) have used this technique with considerable success to study the viscoelastic properties of polymer liquids. The cancellation point can be determined to $\pm 0\cdot1$ dB in amplitude and ± 1 Hz in frequency. This corresponds to an accuracy of about $(\pm 700 \pm i400)$ N s m^{-3} in the liquid shear impedance, and so the viscosity of a liquid of 1 N s m^{-2} viscosity can be determined to about $\pm 3\%$ in the absence of relaxation: in a relaxing liquid, however, the errors are larger. In principle there is no upper limit to the viscosity which may be studied: in practice, the limit is set by the viscosity which may conveniently be poured around the rod. The upper and lower frequency limits are set by the dimensions in which quartz crystals can conveniently be manufactured. This system has an important advantage over the resonant torsional quartz crystal in that it can be driven at frequencies other than the resonant frequency of the crystal because of the heavy damping of the latter. McSkimin (1952) found that the operating range could be made as great as 80% of the midband frequency.

The disadvantages of this system are that it is confined to liquids of relatively high viscosity and fairly large quantities of liquid are required to make a measurement. Moreover, piezoelectric methods of excitation are

not capable of being extended to higher frequencies to fill the gap between
150 kHz and 5 MHz in which no technique is presently available for studying the viscoelastic behaviour of liquids. A modification of this method
has been developed by Glover and coworkers (1968) to overcome these
disadvantages. Torsional oscillations are generated in a hollow tube by a
magnetostrictive method using the Wiedemann effect. A nickel tube is
magnetized circumferentially by passing a current of several hundred
amperes through it for about a millisecond. This tube will twist in proportion to an externally applied axial field from a coil mounted round the tube.
The tube assembly devised by Glover and coworkers (1968) is shown in
Figure 6.7.2. The initial pulse of torsional waves generated by the transmitter coil is propagated along the tube in both directions. The pulse
travelling in an upwards direction is totally absorbed in a cone of grease,
while the other is reflected at the foot of the tube and then activates the
receiver coil. Meanwhile a second transmitted pulse of the same frequency
is generated and arranged to arrive at the receiver coil at the same time as
the reflected pulse. The amplitude of the second pulse and the frequency
are adjusted until the two signals cancel. On applying a liquid to the tube,
the frequency and the amplitude of the second pulse must be adjusted to
restore cancellation, and from these changes the liquid impedance may be
determined. If the tube is wetted within and without with liquid, the
expression for the shear mechanical impedance of the liquid is

$$Z_c = Z_L = R_L + iX_L = \frac{\rho_r U_t a}{2l} \left(\frac{1-n^4}{1+n^3}\right) (\Delta\alpha + i\Delta\beta) \qquad (6.7.3)$$

This is identical to equation (6.7.1) for a solid rod, except for the term
$(1-n^4)/(1+n^3)$, where n is the ratio of the inner to the outer radii of the
tube. For a thin-walled tube where n is near to unity, the measurable
quantities $\Delta\alpha$ and $\Delta\beta$ become comparatively large. This permits a considerable increase in sensitivity over the corresponding system using a solid
rod. Another important advantage of a tube wetted on both sides is that
the cylindrical corrections for the two surfaces are in the opposite sense
and cancel, and the plane wave impedance is measured directly. By altering
the dimensions of the tube and thus the value of n, the sensitivity of the
technique can be arranged to suit the liquid being investigated. Another
advantage is that the tube can be used wetted on one side only. This is
particularly useful when only small quantities of liquid are available:
accurate measurements may be obtained on less than 1 cm³ of liquid contained within the tube. In this case a cylindrical correction must be applied.

Initially magnetostrictive excitation of a tube has been developed to cover
the frequency range 20–200 kHz only, although it is capable of extension
to frequencies of at least 500 kHz. The reproducibility of a measurement is

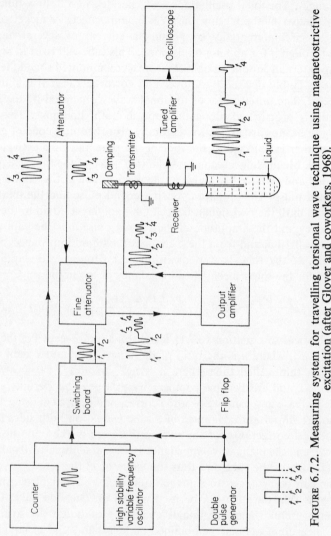

FIGURE 6.7.2. Measuring system for travelling torsional wave technique using magnetostrictive excitation (after Glover and coworkers, 1968).

$(\pm 30 \pm \text{i}30)\,\text{N}\,\text{s}\,\text{m}^{-3}$, and the absolute accuracy $(\pm 150 \pm \text{i}100)\,\text{N}\,\text{s}\,\text{m}^{-3}$. Thus, with suitable calibration against a Newtonian liquid of known viscosity, the components of the shear mechanical impedance of a liquid of $10^{-2}\,\text{N}\,\text{s}\,\text{m}^{-2}$ viscosity may be determined to about 1%, and useful measurements can be made down to viscosities of $10^{-3}\,\text{N}\,\text{s}\,\text{m}^{-2}$. This technique should find considerable application in studying the viscoelastic properties of polymer liquids and solutions.

This magnetostrictive travelling wave technique is not applicable to liquids of very high viscosity: the motion of the tube is too heavily damped and an appreciable part of the energy in the torsional wave is reflected at the liquid surface. Glover and coworkers (1969) have adapted the technique to the high-viscosity region where the liquid behaves elastically at the frequency of measurement and $G'' \approx 0$. A hollow glass tube filled with the liquid to be studied is attached rigidly to the driving, magnetostrictive tube. The glass tube with liquid filling effectively behaves as a solid composite rod along which a single mode of torsional oscillation will travel. The velocity of this wave is determined by the rigidity moduli G' of the core and of the tube: if the latter is known from measurements in air, then G' of the liquid may be found.

The measuring technique for such a composite rod has been described by Bell (1968). The acoustic coupling between the heavy composite rod and the light magnetostrictive tube is weak. Appreciable energy transfer only takes place when the driving frequency coincides with a resonance frequency of the heavy rod which then rings after the driving pulse is removed. From the length and resonant frequency of the composite rod, the torsional wave velocity is obtained.

By combining the composite rod and travelling wave systems in a range of tubes of different sensitivities, the entire liquid viscosity range from 10^{-3} to $10^{12}\,\text{N}\,\text{s}\,\text{m}^{-2}$ may be covered with one technique.

6.8   Normal and Inclined Incidence Techniques: 5–100 MHz

Torsionally vibrating systems cannot be extended to frequencies above 1 MHz, and in this region the shear mechanical impedance of liquids is determined from measurements of the complex reflexion coefficient of a shear wave incident on a solid–liquid interface. Shear waves having frequencies in the megahertz range may conveniently be generated using crystal quartz transducers. If discs are cut from crystalline quartz (Figure 3.1.1) at various orientations and silvered on opposite faces, the discs may execute a shearing motion in their plane when excited by an alternating voltage. A number of different crystal cuts may be used. A Y-cut crystal lies in the plane containing the Z and X axes, while AT-cut and BT-cut crystals are rotated Y-cuts, that is they are rotated about the

X axis to make angles of $35°15'$ and $49°$ with the Z axis respectively. AT-cut crystals are often used to generate shear waves since the dependence of their resonant frequencies on temperature is small.

Mason and McSkimin (1947) have shown that by rigidly bonding a suitably oriented disc of crystal quartz to a fused quartz bar (cf. Section 3.7) and applying a pulse of radio frequency oscillation at a resonant frequency of the disc, very pure shear waves can be propagated along the bar (Figure 6.8.1). If the ends of the bar are accurately flat and parallel, the

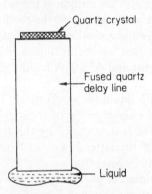

Quartz crystal

Fused quartz delay line

Liquid

FIGURE 6.8.1. Crystal and delay line for normal incidence measurements of the shear mechanical resistance of liquids.

pulse of shear waves is reflected from the other end of the bar and then reactivates the crystal to produce a detectable electrical signal. If the pulse length is less than the transit time of the pulses in the bar, a discrete pulse of shear waves will be reflected up and down the bar many times with ever-decreasing amplitude. When a thin layer of liquid is applied to the end of the bar away from the crystal, the amplitude and phase of the reflected waves are altered, and from these changes the viscoelastic properties of the liquid may be deduced.

On applying the liquid to the bar, the reflexion coefficient for shear waves changes from unity to a value R^*, where R^* is a complex quantity related to the shear impedance of the quartz Z_Q and of the liquid Z_L:

$$R^* = (Z_L - Z_Q)/(Z_L + Z_Q) \qquad (6.8.1)$$

The attenuation of elastic waves in quartz is small at frequencies below 100 MHz, and Z_Q is purely resistive. The attenuation of shear waves in a liquid is considerable, however, and Z_L is complex with resistive and reactive components. If we write

$$R^* = R\exp\left[-i(\pi+\theta)\right] \tag{6.8.2}$$

where R is the absolute value of the reflexion coefficient and $(\pi + \theta)$ is the phase change on reflexion, we can rearrange equation (6.8.1) to give

$$Z_L = R_L + iX_L = Z_Q\left(\frac{1-R^2+i2R\sin\theta}{1+R^2+2R\cos\theta}\right) \tag{6.8.3}$$

Where a shear wave is reflected from a quartz–air interface, $R = 1$ and $\theta = 0$. When a liquid is applied to the quartz surface, R is reduced and θ is increased. Hence if the changes in the amplitude and phase of the reflected wave can be measured, the components of the shear mechanical impedance of the liquid can be calculated from equation (6.8.3) using known values of Z_Q. The attenuation of shear waves in the liquid is so large that with a relatively thin layer of liquid the amplitude and phase changes are independent of liquid thickness.

Since the limiting shear rigidity G_∞ of the liquid is much smaller than the rigidity of the solid, even the maximum attainable value of Z_L $[=(\rho G_\infty)^{\frac{1}{2}}]$ is much smaller than Z_Q. Hence the value of R^* is close to unity for the normal incidence technique, and the associated phase and amplitude changes are very small. An increase in the sensitivity of the technique may be obtained by reflecting the shear waves at a large angle to a plane surface (Mason and coworkers, 1949). A suitable device for this inclined incidence technique is shown in Figure 6.8.2. The quartz crystal must be

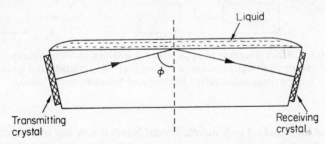

FIGURE 6.8.2. Crystals and delay line for inclined incidence measurements of the shear mechanical impedance of liquids.

oriented so that the particle motion at the reflecting surface is normal to the direction of propagation and parallel to the surface: this gives a pure shearing motion at the interface, and does not produce a longitudinal component on reflexion. O'Neil (1949) has shown that under these

5

conditions the liquid impedance is

$$Z_L = R_L + iX_L = Z_Q\left(\frac{1 - R^2 + i2R\sin\theta}{1 + R^2 + 2R\cos\theta}\right)\cos\phi \qquad (6.8.4)$$

where ϕ is the angle of incidence of the shear wave on the surface. Thus in principle a value of ϕ approaching 90° would lead to substantial changes of amplitude and phase. This would require an excessively large bar, however, so that in practice a compromise is adopted with $\cos\phi$ of the order of 0·25.

Figure 6.8.3 is a block diagram of the circuitry used by Barlow and Subramanian (1966) in their development of the inclined incidence technique. The output of a continuous wave, high stability oscillator is gated,

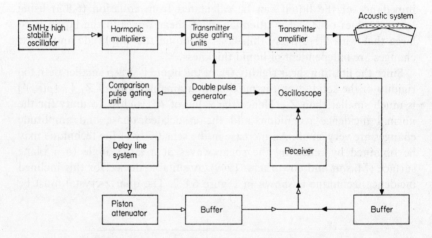

FIGURE 6.8.3. Block diagram of apparatus for inclined incidence measurements of the shear mechanical impedance of liquids: —▶—, transmitted pulse; —▷—, comparison pulse (after Barlow and Subramanian, 1966).

amplified and applied to a quartz crystal bonded to a bar of fused quartz. The resultant pulse of shear waves is reflected to and fro within the quartz bar, and at each reflexion at the quartz crystal part of the elastic energy is reconverted into an electrical signal. The train of received pulses is amplified, demodulated and displayed on an oscilloscope. Its amplitude is compared with that of a delayed pulse derived from the same oscillator: this pulse has passed through a precision attenuator, and can be adjusted to have the same amplitude as the pulse received from the bar. When a liquid is applied to the bar, the change in the piston attenuator setting corresponds to the decrease in amplitude of the shear waves as a result of their reflexions at the quartz–liquid interface. For the phase measurement the reference signal is altered in phase by a precision delay line: the reference signal is super-imposed on a received signal, and its amplitude and phase adjusted until it exactly cancels the main signal. On applying the liquid, cancellation is restored by altering the delay line and attenuator settings, and X_L may be calculated from the difference in the two delay line settings. It is customary to measure the amplitude and phase changes over several reflexions to improve accuracy. In this way Barlow and Subramanian (1966) were able to determine R_L and X_L of a liquid to $\pm 3 \times 10^3$ and $\pm 5 \times 10^3 \, \mathrm{N s m^{-3}}$ respectively at 5 MHz: at 75 MHz, the error in X_L is increased to $\pm 10^4 \, \mathrm{N s m^{-3}}$.

A method of determining phase changes with a normal incidence system has been described by McSkimin and Andreatch (1967). This is a pulse superposition method in which the pulse repetition frequency is chosen so that the time between applied pulses is approximately twice the time required for shear waves to make a complete round trip in the bar. By altering the repetition rate by a very small amount, the amplitude of the superimposed odd-ordered echoes can be maximized showing that the signals are exactly in phase. On applying the liquid to the bar, the phase change undergone by the shear waves on reflexion can be calculated from the change in the pulse repetition frequency required to return the super-imposed echoes to maximum amplitude. Because of the frequency resolution attainable (1 in 10^7 at 40 MHz) McSkimin and Andreatch (1967) were able to measure X_L to $\pm 2 \times 10^3 \, \mathrm{N s m^{-3}}$.

In order to achieve this accuracy in X_L, however, it is necessary that the temperature of the quartz bar be held constant to a few millidegrees. Should this stability not be attainable, it is more convenient to dispense with the phase measurement channel and measure amplitude changes only with the normal incidence system. Since $Z_L < 0{\cdot}1(1+\mathrm{i})Z_Q$, the phase angle θ is less than $5°$, and equation (6.8.3) may be written

$$R_L \approx Z_Q(1 - R)/(1 + R) \tag{6.8.5}$$

The normal incidence technique is particularly useful for studying liquids at frequencies above the viscoelastic relaxation region. Here the liquid response is purely elastic, $X_L = 0$ and $G_\infty = R_L^2/\rho$. Normal incidence measurements have been widely used to study the variation of G_∞ with temperature (Barlow and coworkers, 1967).

In liquids of viscosity greater than $10^3 \, \mathrm{N s m^{-2}}$, the attenuation of shear waves becomes sufficiently small for them to be propagated an appreciable distance through the liquid. An alternative to the normal incidence tech-nique is then to measure the energy transmitted through the liquid. This method is identical to the pulse technique for studying the propagation of longitudinal waves described in Section 3.7: an example of the use of such a system with shear waves is given by Simmons and Macedo (1968).

6.9 Normal Incidence Technique: 300–1600 MHz

At frequencies above 100 MHz, the absorption of elastic waves in quartz delay lines becomes appreciable. It is also difficult to achieve good coupling between the transducer and the delay line (cf. Section 3.7). Hence at higher frequencies it becomes preferable to generate shear waves directly in a piezoelectric crystal (cf. Section 3.1). In the technique of Bömmel and Dransfeld (1960), pulses of radio-frequency oscillation excite an electro-magnetic cavity at its resonant frequency. One end of a suitably oriented rod of crystal quartz is placed in the cavity, and the oscillating electric field in the cavity generates a train of elastic waves at the surface of the quartz. These waves are reflected from each end of the rod, and at each reflexion at the cavity a small proportion of the elastic wave re-excites the cavity and produces an observable signal. In this way it is possible to record the decay of the elastic energy in the rod. If a liquid is applied to the end of the rod outside the cavity, the shear waves experience a change of phase and amplitude on reflexion at the quartz–liquid interface. The components of the shear mechanical impedance of the liquid can then be deduced from these changes as in the lower frequency technique of the previous section.

Figure 6.9.1 is a simplified diagram of the type of cavity devised by Lamb and Richter (1966) for studying the viscoelastic properties of liquids.

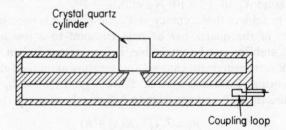

FIGURE 6.9.1. Simplified diagram of cavity resonator for generating shear waves at micro-wave frequencies (after Lamb and Richter, 1966).

It consists of two adjustable, short-circuited, coaxial lines, each having a length approximately equal to a quarter wavelength. The lines are coupled through the gap capacitance and are tunable from 300 to 1600 MHz in the fundamental mode. When a pulse of radio-frequency oscillation is applied to this cavity with the outer end of the quartz crystal in air, over fifty echoes can be observed at room temperature for the propagation of 500 MHz shear waves in a rod of crystal quartz 15 mm long, the r.m.s. deviation of the pulses from an exponential decay rate being 0·05 dB.

At present it is only possible to measure the change in amplitude of the shear wave pulses in the quartz on application of a liquid. The experimental method of doing this is similar to that in the lower frequency normal incidence technique (Section 6.8): a reference pulse of r.f. oscillation can be adjusted by a calibrated attenuator and is compared with the amplitude of the pulses received from the cavity. From the additional attenuation of the shear waves caused by the liquid, the resistive part of the shear mechanical impedance of the liquid can be determined to better than $\pm 5 \times 10^3$ N s m^{-3}. This technique has been successfully used by Barlow and coworkers (1967) to study the viscoelastic behaviour of a number of liquids at frequencies between 300 and 1600 MHz.

6.10 Normal Incidence Technique: > 1 GHz

Since the attenuation of elastic waves in crystal quartz increases as the square of the frequency, an alternative to the previous technique is desirable, combining higher transmitted power with a shorter acoustic path length. Such a technique has been devised by Lamb and Seguin (1966) at

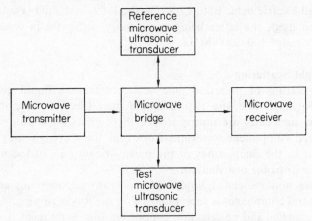

FIGURE 6.10.1. Block diagram of apparatus for measuring the shear mechanical resistance of liquids at 3 GHz (after Lamb and Seguin, 1966).

3 GHz, and their system is shown schematically in Figure 6.10.1. Pulses of 250 kW peak power from a transmitter are coupled through a microwave bridge into two tuned cavities (cf. Section 6.9). Each cavity contains a cylindrical crystal of quartz 3 mm thick in which shear waves are generated. On reflexion at the ends of the quartz crystals, the shear waves

re-energize the cavities and produce a detectable signal: the response from the two identical cavities is balanced on the microwave bridge. When a drop of liquid is applied to one crystal, it is necessary to insert attenuation and alter the phase in the other arm of the bridge to re-establish cancellation. Hence in principle it is possible to calculate both components of the shear mechanical impedance of the applied liquid.

In practice the reproducibility of the phase measurements was inadequate to permit the determination of the reactive part of the liquid shear impedance, X_L. The results for R_L, however, are in reasonable agreement with the predictions from measurements at lower frequencies.

The conversion of electrical to acoustic energy and the subsequent reconversion to electrical energy are so inefficient that the accurate determination of the small amplitude change on addition of the liquid presents a formidable experimental problem at gigahertz frequencies. The situation is greatly improved by the use of thin film evaporated transducers (Section 3.1). For studies of liquids these are deposited on a suitable delay line, while they can be evaporated directly onto the solid to be investigated (cf. Section 3.7). Under certain conditions vapour deposition at oblique incidence can align the crystal axes of a semiconducting film in such a way as to cause the film to execute a shearing motion under the influence of an applied electric field. Both CdS (Curtis, 1969) and ZnO (Foster, 1969) have been used, the latter being particularly successful in covering the frequency range 100–3000 MHz.

6.11 Light Scattering

The spectrum of polarized light scattered from a liquid contains a central Rayleigh line and two Brillouin lines (Section 3.8). In liquids where the molecules are non-spherical, there exists an additional broad depolarized component sometimes known as the Rayleigh wing. This is attributed to the fluctuations of the mean orientations of the molecules from their isotropic distribution.

In some non-associated liquids of moderate viscosity, an additional sharp central component is superimposed on the Rayleigh wing. At certain scattering angles and polarization conditions, this component is split into two peaks symmetrically disposed about the incident light frequency. This doublet arises from a coupling between the orientations of the molecules and high frequency shear waves in the liquid (Starunov, Tiganov and Fabelinskii, 1966). From the detailed shape of this doublet, the velocity of shear waves, the rigidity modulus G_∞ and the shear relaxation time of the liquid may be deduced (Volterra, 1969), and some experimental results are available (Stegeman and Stoicheff, 1968).

CHAPTER 7

THE VISCOELASTIC PROPERTIES OF DILUTE POLYMER SOLUTIONS

7.1 Polymer Molecules in Solution

In solution a flexible polymer molecule is coiled randomly, the overall size of the coil being determined by the polymer–solvent interactions. In a thermodynamically 'good' solvent, where polymer–solvent contacts are favoured, the coils are relatively extended, while in a 'poor' solvent the coils are more compact. Polymer molecules are in constant thermal motion among a large number of conformations of equal energy. It is convenient to describe the dimensions of a polymer molecule in solution by the root-mean-square separation of the ends of the molecule. Provided that the molecules are freely orienting, with no fixed bond angles and no restriction of intramolecular rotation, the probability $P(x, y, z) \, dx \, dy \, dz$ of finding one end of the molecule in the volume element of size $dx \, dy \, dz$ located at (x, y, z) when the other end is at the origin of the coordinate system is

$$P(x, y, z) \, dx \, dy \, dz = \left(\frac{3}{2\pi Nl^2}\right)^{\frac{3}{2}} \exp\left[-\left(\frac{3}{2Nl^2}\right)(x^2 + y^2 + z^2)\right] dx \, dy \, dz \quad (7.1.1)$$

Here N is the number of chain segments of length l. This is a Gaussian distribution, and it is applicable provided that the chain is not extended near to its maximum length.

Equation (7.1.1) implies that the average vector distance between the chain ends of a polymer molecule in solution is zero. If we clamp the ends of the molecule in any two positions and try to pull the ends apart, we are forcing the molecule to assume a less probable conformation: when the molecule is released, it will on average return to a more compact conformation. Thus the polymer molecule is behaving as an entropy spring. When a polymer solution is subjected to a shearing force, the resulting velocity gradient in the solvent disturbs the equilibrium conformation of a polymer molecule. The subsequent motion of the polymer segments through the solvent during the return to equilibrium leads to viscous energy losses. This segmental motion can be considered to result from a series of normal modes of motion of the polymer chain as shown diagrammatically in Figure 7.1.1.

The rate of approach of a mode towards its equilibrium conformation is characterized by a relaxation time τ. At low frequencies of alternating strain and at low rates of steady shear, each mode has ample time to respond to the applied strain and hence to contribute to the steady flow viscosity of the solution. At higher frequencies of alternating strain or at higher rates of shear there is insufficient time for all modes to adjust themselves. The solution shows viscoelastic behaviour: the dynamic viscosity falls while the rigidity modulus becomes finite (cf. Chapter 5).

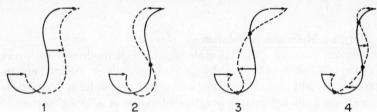

FIGURE 7.1.1. Schematic representation of the first four normal modes of a polymer molecule.

Provided that the distribution of end-to-end distances of each mode of motion follows a Gaussian distribution function, the contribution of each mode to the viscosity and rigidity of the solution may be calculated from a number of molecular theories as discussed in the next section.

7.2   Molecular Theories of the Viscoelastic Properties of Dilute Polymer Solutions

In the theory of Rouse (1953) a polymer molecule is divided into N equal segments, each segment containing sufficient monomer units (say 20) for the equilibrium separation of the ends of each segment to be given by the Gaussian probability function (equation 7.1.1). Each segment acts as an entropy spring having a finite rigidity, and the relative motion of segment and solvent gives rise to viscous losses. When a shear stress is applied to a polymer solution, the conformations of the segments are no longer distributed randomly, and the thermal motions act to restore this random conformation. Rouse used the theory of rubber elasticity to calculate the entropy change resulting from such a disturbance of a polymer segment. For small perturbations all conformations of the polymer molecule are of equal energy, and so the change in the free energy and the chemical potential of the molecules depend only on the entropy change. The velocity with which a segment returns to its equilibrium conformation is assumed to be given by the product of the mobility (i.e. the ratio of restoring force to the segmental velocity) of the end of a polymer segment and the change in the chemical potential of the segment resulting from its displacement. The segmental mobility depends upon the viscous drag exerted by the solvent on a moving segment. Rouse assumed that this drag is concentrated at the junctions of the segments and that all segments of the molecule experience the same viscous drag irrespective of their position in the molecule.

The set of N partial differential equations which describe the components of the motion of the polymer molecules can be written as a matrix equation, the eigenvalues of which are related to the mode number p by the relation

$$\lambda_p = 4\sin^2[p\pi/2(N+1)] \qquad (7.2.1)$$

Here λ_p is the pth of the set of N eigenvalues. By means of an orthogonal transformation of coordinates, the characteristic relaxation time for the return of each polymer mode to its equilibrium conformation may be found in terms of these eigenvalues:

$$\tau_p = \sigma^2/6BkT\lambda_p \qquad (7.2.2)$$

Here σ is the root-mean-square separation of the ends of a polymer segment, and B is the segmental mobility.

In order to obtain the contribution of each mode of motion of the polymer molecules to the complex shear modulus of the solution, Rouse equated the power supplied by the shear waves with the rate at which work is done on the solvent and on the polymer molecules. The latter may be calculated from the rate of increase in the free energy of the polymer molecules which results from the motion of each segment of the molecule with the same velocity as the adjacent solvent molecules. From this Rouse derived expressions for the rigidity and loss moduli of the polymer solution in terms of the cyclic frequency ω, the solvent velocity η_S, and the number of polymer molecules in unit volume n:

$$G' = nkT \sum_{p=1}^{N} \frac{\omega^2 \tau_p^2}{1+\omega^2 \tau_p^2} \qquad (7.2.3)$$

$$G'' = \omega\eta' = \omega\eta_S + nkT \sum_{p=1}^{N} \frac{\omega\tau_p}{1+\omega^2 \tau_p^2} \qquad (7.2.4)$$

Equations (7.2.3) and (7.2.4) are generalizations of the Maxwell model of Section 5.4 in which each Maxwell element has a value of nkT for G_∞. This entropy modulus increases with temperature in contrast to the energy modulus in simple liquids discussed in Section 9.3.

The values of the relaxation times τ_p for the motions of the various modes of the polymer molecules are obtained by combining equations (7.2.1) and (7.2.2):

$$\tau_p = \sigma^2 \left[24BkT\sin^2\frac{p\pi}{2(N+1)} \right]^{-1}, \quad p = 1, 2, ..., N \qquad (7.2.5)$$

When $p < N/5$, equation (7.2.5) simplifies to

$$\tau_p \approx \frac{\sigma^2(N+1)^2}{6\pi^2 p^2 BkT} \tag{7.2.6}$$

Neither σ, N nor B may be calculated directly. However at frequencies below the polymer relaxation region ($\omega\tau \ll 1$) where the dynamic viscosity η' of the solution equals the steady flow viscosity η_s, equation (7.2.4) may be written

$$\eta_s = \eta_S + nkT \sum_{p=1}^{N} \tau_p \tag{7.2.7}$$

$$= \eta_S + \frac{n\sigma^2}{24B} \sum_{p=1}^{N} \left[\sin\frac{p\pi}{2(N+1)}\right]^{-2}$$

$$= \eta_S + \frac{n\sigma^2 N(N+2)}{36B}$$

$$\approx \eta_S + \frac{n\sigma^2(N+1)^2}{36B} \tag{7.2.8}$$

Combination of equations (7.2.6) and (7.2.8) then yields an expression for τ_p in terms of the steady flow viscosities of solution η_s and solvent η_S:

$$\tau_p = \frac{6(\eta_s - \eta_S)}{\pi^2 p^2 nkT}, \quad p = 1, 2, ..., N \tag{7.2.9}$$

Equations (7.2.3), (7.2.4) and (7.2.9) are the predictions of the Rouse theory for the variation with the frequency of alternating strain of the rigidity and loss moduli of a dilute polymer solution. G' and G'' are expressed in terms of the steady flow viscosity of the solvent and solution, and the number of polymer molecules in unit volume of solution: there are no adjustable parameters. The results are shown in Figure 7.2.1.

An alternative theory to that of Rouse was put forward by Bueche in 1954. This uses somewhat different assumptions to derive a result closely similar to that of Rouse. The model used in deriving such theories breaks down at large values of the mode number p where the length of a chain segment is too small for the distribution of segment ends to follow the Gaussian probability function. The Rouse model assumes that there is no hydrodynamic interaction between the motions of the segment junctions, that is, the polymer chain is assumed to be free-draining. The solvent velocity is not affected by the presence of the molecule, and the viscous drag experienced by a polymer segment in moving through the solvent is the same for all segments irrespective of their position in the molecule.

Moreover, intramolecular steric effects are neglected, and it is assumed that the junctions between chain segments consist of perfectly free, universal joints.

In the Zimm (1956) theory of the viscoelastic properties of dilute polymer solutions, the effect of hydrodynamic interaction between segments of the polymer chain is explicitly considered. The presence of a segment of a polymer chain distorts the flow pattern of the solvent

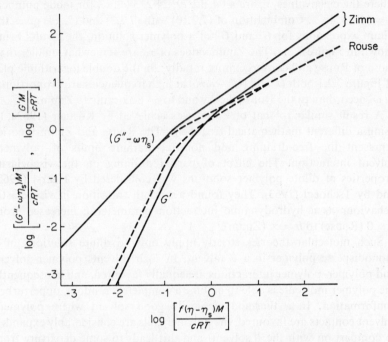

FIGURE 7.2.1. Contribution of a polymer solute to the components of the complex shear modulus as predicted by the theories of Zimm (—) and Rouse (----).

through the molecule, i.e. the polymer molecule is non-free-draining. The exact mathematical description of this would be complex since the flow lines of the solvent are distorted into a curvilinear pattern by the chain segments. Zimm assumed that the Kirkwood–Riseman (1948) approximation, an extreme case of dominant interaction, could be applied and that the curvilinear flow field could be approximated by a rectilinear pattern in the neighbourhood of a moving segment. Zimm was then able to calculate the velocity of the flowing solvent at each segment of the polymer chain.

The analysis of the motion of the segments in response to the solvent flowing in this way is similar to that of Rouse. The resulting values of the relaxation times for the modes of motion of a non-free-draining polymer chain are

$$\tau_p = \frac{1 \cdot 71(\eta_s - \eta_S)}{\lambda'_p \, nkT} \qquad (7.2.10)$$

where the eigenvalues λ'_p are 4·04, 12·79, 24·2, 37·9, ... for mode numbers $p = 1, 2, 3, 4, ...$ Combination of (7.2.10) with (7.2.3) and (7.2.4) gives the Zimm expressions for G' and G'' of a polymer solution, the results being shown in Figure 7.2.1. The Zimm values of τ_p are somewhat smaller than those of Rouse and converge more rapidly: on the double logarithmic plot of Figure 7.2.1 both G' and $(G'' - \omega\eta_S)$ at high frequencies are proportional to $\omega^{\frac{1}{2}}$ according to the Rouse theory and to $\omega^{\frac{2}{3}}$ according to Zimm.

A result similar to that of Zimm was achieved by Kästner (1962a, b) using a different mathematical treatment. The Rouse and Zimm theories represent the free-draining and non-free-draining limits of polymer–solvent interaction. The effect of partial draining on the viscoelastic properties of dilute polymer solutions was considered by Hearst (1962) and by Tschoegl (1963). They found a smooth transition in viscoelastic behaviour as a hydrodynamic interaction parameter h increased from $h = 0$ (Rouse) to $h = \infty$ (Zimm).

Such molecular theories strictly apply only to dilute solutions of a monodisperse polymer in a θ solvent. In such solvents polymer–solvent and polymer–polymer interactions are equally favoured, and consequently the polymer molecule is able to assume a completely random, unperturbed conformation. In a thermodynamically good solvent where polymer–solvent contacts are favoured, the polymer coils are considerably expanded in comparison with the θ solvent, and this leads to some departure from Gaussian chain statistics. Ptitsyn and Eisner (1958, 1959) modified the Zimm theory to allow for this effect, but it transpired that the alterations required are small. A particularly general treatment of the viscoelastic properties of dilute polymer solutions is that of Tschoegl (1964) where the effects of varying hydrodynamic interaction and solvent power are considered. The Rouse and Zimm theories have also been extended to polymers having a distribution of molecular weights (Lovell and Ferry, 1961).

The molecular theories already discussed assume that all of the polymer contribution to the solution viscosity comes from the modes of motion of the polymer chain as expressed by equation (7.2.4). This implies that the dynamic viscosity of a polymer solution should approach the solvent viscosity at frequencies of alternating strain above the polymer relaxation region. Experimentally Lamb and Matheson (1964) and Philippoff (1964) have found that the high frequency limiting value of the dynamic viscosity η_∞ is significantly higher than the solvent viscosity. Lamb and Matheson suggested that this effect arose from the finite volume of a polymer molecule by analogy with the Einstein (1906, 1911) equation for the viscosity of a suspension of rigid spheres.

The Rouse and Zimm theories assume that the junctions between the polymer segments are perfectly free universal joints and that the polymer molecule can respond instantaneously to an applied force. In a real polymer molecule, however, changes of shape are the result of local conformational changes by internal molecular rotation about the bonds in the polymer chain. The finite barriers to such conformational changes (Chapter 12) necessitate a finite time delay before the conformational change occurs, and this leads to a viscosity type relationship between the applied force and the resulting deformation rate. Cerf and Thurston (1964) suggested that as a result of this internal viscosity of the polymer chain, the distortion of the polymer molecule by the flowing solvent differs from that assumed in the Rouse and Zimm models.

Peterlin (1967) and Peterlin and Reinhold (1967) included this internal viscosity in an analysis of the motion of the polymer chain, and derived the following values of the components of the rigidity modulus of the solution:

$$G' = \omega\eta'' = nkT \sum_{p=1}^{N} \frac{\omega^2 \tau_p^2}{1+\omega^2 \tau_p'^2} \tag{7.2.11}$$

$$G'' = \omega\eta' = \omega\eta_S + nkT \sum_{p=1}^{N} \frac{\omega\tau_p[1+\omega^2\tau_p'(\tau_p'-\tau_p)]}{1+\omega^2\tau_p'^2} \tag{7.2.12}$$

Here

$$\tau_p' = (1+p\nu_p\phi/N\zeta)\tau_p \tag{7.2.13}$$

where ν_p are the eigenvalues of the tensor of internal viscosity, ϕ is the coefficient of internal viscosity and ζ is the segmental friction coefficient or force required to pull the segment through its surroundings at unit speed. The values of τ_p are given by the theories of Rouse (1953), Zimm (1956) or Tschoegl (1964) for the appropriate degree of hydrodynamic interaction. In the case of a free-draining coil, $\nu_p = 1$, and the components of the polymer contribution to the dynamic solution viscosity are shown in Figure 7.2.2 for various values of $a = \phi/N\zeta$.

At low frequencies ($\omega\tau_1 < 1$) the curves for η' are not affected by the presence of the internal viscosity, and superimpose on the Rouse curve ($a = 0$) for a completely flexible coil. At high frequencies ($\omega\tau_1 > 1$), however, the internal viscosity curves deviate from the Rouse theory and attain a limiting value η_∞ which is greater than the solvent viscosity. The curves for η'' for the free-draining coil become flattened and attain a maximum at lower frequencies as the internal viscosity increases. The Peterlin (1967) predictions for the non-free-draining case are similar. Since the coefficient of internal viscosity is not easily obtained experimentally, the Peterlin theory contains an adjustable constant which is not present in theories which neglect internal viscosity. Nevertheless the prediction of η_∞ represents a significant advance over earlier theories.

The viscoelastic properties of dilute solutions of rigid, rod-like polymer molecules are entirely different from those of solutions of flexible molecules.

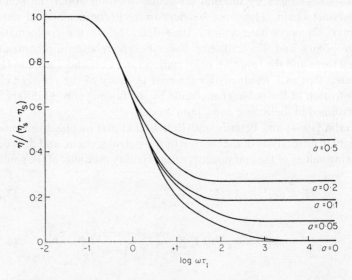

FIGURE 7.2.2. Frequency dependence of the contribution of a polymer solute to the normalized real part of the dynamic viscosity of a polymer solution for a free-draining coil with different values of the internal viscosity. $a = \phi/N\zeta$ where ϕ is the coefficient of internal viscosity, Z is the number of segments in the polymer chain, and ζ is the friction coefficient of a segment. τ_1 is calculated from equation (7.2.9) (after Peterlin, 1967).

A rigid molecule in solution is caused to rotate by an applied velocity gradient, the resultant relative motion of polymer molecule and solvent leading to viscous energy losses. When the frequency of the applied strain becomes comparable to the relaxation time for this rotational motion, then an elastic modulus is observed. Kirkwood and Auer (1951) considered the rotary motion of a rigid rod-like polymer molecule in solution, and showed that a single relaxation time is sufficient to describe the viscoelastic behaviour:

$$G' = \frac{3nkT}{5} \frac{\omega^2 \tau^2}{1+\omega^2 \tau^2} \tag{7.2.14}$$

$$G'' - \omega\eta_S = \frac{3nkT\omega\tau}{5} \left(\frac{1}{1+\omega^2 \tau^2} + \frac{1}{3} \right) \tag{7.2.15}$$

where

$$\tau = \frac{\pi\eta_S L^3}{18kT} \ln \frac{L}{b} \tag{7.2.16}$$

L is the length of the molecule, b the length of its repeat unit and η_S the solvent viscosity. A similar result has been obtained by Cerf (1952) and by Scheraga (1955) for solutions of rigid ellipsoidal molecules. The results of these calculations are shown in Figure 7.2.3.

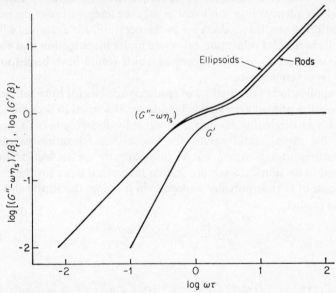

FIGURE 7.2.3. Contribution of a polymer solute to the components of the complex shear modulus as predicted by the theories of Kirkwood and Auer (1951) for rods, and Cerf (1952) and Scheraga (1955) for ellipsoids.

Many cases intermediate between rigid rods and completely flexible chains can be envisaged. The properties of solutions of such molecules are often discussed in terms of the Kratky–Porod (1949) worm-like or semi-flexible chain, or in refinements of this such as the thin flexible wire model of Bugl and Fujita (1969). The Kratky–Porod model chain follows a continuous smooth curve which changes direction at random. Its stiffness is characterized by a persistence length which represents the distance in one direction in which a given property of the molecule is unchanged. Harris and Hearst (1966) have introduced a stiffness parameter into the calculation of the dynamic properties of a polymer chain, and have related the components of the complex rigidity modulus of the solution to this stiffness.

7.3 The Method of Reduced Variables

Since viscoelastic relaxation in polymer solutions can occur over many decades of frequency, it is desirable to have a measuring technique which will cover an equally wide frequency range. Most of the experimental systems discussed in Chapter 5 can only operate successfully at a fixed frequency or over a small range of frequencies. In order to extend the available frequency range it is usual to vary the temperature of the polymer solution being studied: a decrease in temperature, for example, will slow down the molecular relaxation processes under investigation and enable a given instrument to observe processes which would have been too rapid at the higher temperature.

The application of the method of reduced variables (or time–temperature superposition principle) to simple liquids was discussed in Section 5.7. The method was originally developed empirically for polymers in order to reduce the experimental results for viscoelastic relaxation at different temperatures to a common curve. Subsequently the molecular theories discussed in Section 7.2 have provided a theoretical basis for the method. In the case of flexible polymer molecules in solution, the Rouse and Zimm theories predict

$$\tau_p \propto \frac{(\eta_s - \eta_S)}{nkT} \propto \frac{M(\eta_s - \eta_S)}{\rho RT} \qquad (7.3.1)$$

and

$$G' \text{ and } (G'' - \omega\eta_S) \propto nkT \propto \rho RT/M \qquad (7.3.2)$$

where ρ is the solution density at temperature T. If an arbitrary reference temperature T_r is chosen, then measurements at any other temperature may be reduced to the reference temperature for a given solution:

$$(G')_r = G'T_r \rho_r/T\rho = G'/\beta_r \qquad (7.3.3)$$

$$(G'' - \omega\eta_S)_r = (G'' - \omega\eta_S) T_r \rho_r/T\rho$$

$$= (G'' - \omega\eta_S)/\beta_r \qquad (7.3.4)$$

while the reduced frequency of measurement f_r is related to the actual frequency f by

$$f_r = f(\eta_s - \eta_S) T_r \rho_r / (\eta_s - \eta_S)_r T\rho = f a_T \qquad (7.3.5)$$

The most pronounced effect of a change of temperature is on $(\eta_s - \eta_S)$. Provided that no structural changes occur in the solution, a modest change in temperature permits the study of a wide range of relaxation processes. The validity of this method has been demonstrated experimentally (Ferry, 1961). It has also been extended to include the effect of pressure changes (Philippoff, 1963).

7.4   Experimental Results for the Viscoelastic Properties of Dilute Polymer Solutions

The first studies of the viscoelastic properties of polymer solutions at concentrations sufficiently low for intermolecular interactions to be negligible were those of Baker, Mason and Heiss in 1952 using the resonant torsional quartz crystal (Section 6.6). They studied solutions of polystyrene and polyisobutylene in benzene and cyclohexane and observed a decrease in the dynamic viscosity and an increase in the rigidity modulus of the solution with increasing frequency. Subsequently Rouse and Sittel (1953) investigated dilute solutions of polystyrene fractions in a number of hydrocarbon solvents over a range of frequencies from 200 Hz to 60 kHz using the torsion pendulum (Section 6.5) and the torsional crystal. Figure 7.4.1 shows the results of Rouse and Sittel (1953) for the dynamic

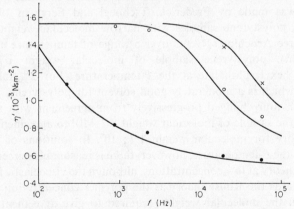

FIGURE 7.4.1. Results of Rouse and Sittel (1953) for the real part of the dynamic viscosity of solutions of polystyrene in toluene at 303·5 K. x, $\bar{M}_w$ $2\cdot53 \times 10^5$, $14\cdot8$ g l^{-1}, η_{rel} $2\cdot99$; $\circ$, $\bar{M}_w$ $5\cdot2 \times 10^5$, $8\cdot9$ g l^{-1}, η_{rel} $3\cdot00$; $\bullet$, $\bar{M}_w$ $6\cdot2 \times 10^6$, $1\cdot44$ g l^{-1}, η_{rel} $3\cdot12$; —, predictions of Rouse theory.

viscosity of three polystyrene fractions in toluene at 303 K. These results show reasonable agreement with the Rouse theory, demonstrating for the first time that the motion of segments of the polymer chain is responsible for the viscoelastic behaviour of polymers.

With the advent of monodisperse polymer samples, Ferry and his collaborators were able to carry out a more detailed comparison of the various molecular theories with experiment. Since their measurements were made using the apparatus of Birnboin and Ferry (1961) (Section 6.4), solvents of high viscosity were required to bring the polymer relaxation times into the experimentally accessible region below 400 Hz. In 1962 De Mallie and coworkers studied solutions of polystyrene with $\bar{M}_w = 2 \cdot 67 \times 10^5$ and a sharp molecular weight distribution in Arochlor 1248, a chlorinated diphenyl, as solvent: the concentrations used were between 0·5 and 4% and the temperature range was 270–320 K. Figure 7.4.2 is a plot reduced to 298 K by the method of reduced variables of the experimental results for a 1% solution, together with the predictions of the Zimm theory. It can be seen that the results are in good agreement with the Zimm theory. This agreement is obtained, however, only if it is assumed that the molecular weight of the polymer is larger than its true value: for this polystyrene sample of molecular weight $2 \cdot 67 \times 10^5$, the molecular weight required to give agreement with the Zimm theory increases from $2 \cdot 75 \times 10^5$ to $6 \cdot 46 \times 10^5$ as the concentration of polymer in solution increases.

An exhaustive study of the viscoelastic properties of dilute polystyrene solutions was made by Frederick, Tschoegl and Ferry in 1964. They studied six polystyrene samples with narrow molecular weight distributions in three Arochlor solvents over a range of temperature and also a monodisperse polystyrene sample of molecular weight $1 \cdot 7 \times 10^6$ in di-(2-ethyl hexyl) phthalate at the θ-temperature. For 2% solutions in Arochlor, which is a moderately good solvent for polystyrene, the viscoelastic behaviour changed progressively from agreement with the Zimm theory for the sample of molecular weight $8 \cdot 2 \times 10^4$ to agreement with the Rouse theory for molecular weight $1 \cdot 7 \times 10^6$. In solutions of the latter polymer at the θ-temperature, however, there is reasonable agreement with the Zimm theory at low concentrations, although the viscoelastic behaviour moves towards the Rouse theory as the polymer concentration increases. Once again the molecular weights required to give agreement with the molecular theories are larger than the true molecular weight, values up to 6·8 times as great being necessary. Similar conclusions have been reached for dilute polyisobutylene solutions in a naphthenic oil (Tschoegl and Ferry, 1964a). Low molecular weight fractions obey the Zimm theory, while solutions of a polymer fraction of a viscosity average molecular weight of $8\cdot4 \times 10^5$ show deviations from the Zimm theory in the direction of Rouse: in all cases the apparent molecular weight is about twice the correct value. Likewise solutions of anionically prepared poly-α-methyl styrene of molecular weights $3\cdot49 \times 10^5$ and $6\cdot3 \times 10^5$ obeyed the Zimm

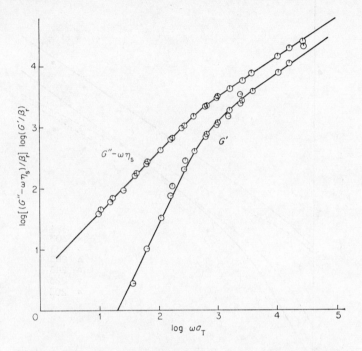

FIGURE 7.4.2. Results of De Mallie and coworkers (1962) for G' and $(G'' - \omega\eta_s)$ for a 1% solution of polystyrene in pentachlorodiphenyl: $\bar{M}_w$, $2\cdot67 \times 10^5$; $\bar{M}_w/\bar{M}_n$, $1\cdot08$; results reduced to $298\cdot2$ K; —, predictions of Zimm theory using $\bar{M}_w = 2\cdot75 \times 10^5$; ⊘, $282\cdot6$ K; ⊖, $298\cdot0$ K.

theory at low concentrations but with discordant molecular weights (Frederick and Ferry, 1965).

From studies of the viscoelastic properties of dilute polystyrene solutions in low viscosity solvents using the resonant torsional crystal, Lamb and Matheson (1964) came to somewhat different conclusions from those of Ferry and his associates. In dilute solutions of low molecular weight polystyrene in the good solvent toluene, there was reasonable agreement with the Zimm theory using the correct molecular weight of the polymer. But deviations from the Zimm theory were observed as the molecular weight of the polymer increased, as the concentration of the polymer increased or as the solvent became thermodynamically poorer. Figure 7.4.3 is a plot of the results for a 1% solution of polystyrene of

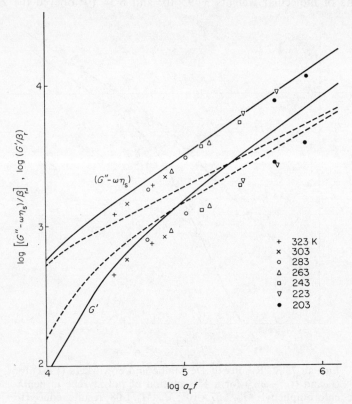

FIGURE 7.4.3. Results of Lamb and Matheson (1964) for G' and $(G'' - \omega\eta_S)$ for a 1·15% solution of polystyrene in toluene: $\bar{M}_w = 3·65 \times 10^5$; $\bar{M}_w/\bar{M}_n = 1·08$; results reduced to 303·2 K; ----, predictions of the Rouse theory; —, predictions of the Zimm theory.

$\bar{M}_w = 3·64 \times 10^5$ in toluene: although the experimental results agree with the Zimm predictions of $(G'' - \omega\eta_S)$ within experimental error, the results for G' are not in agreement with either the Rouse or Zimm theories.

Lamb and Matheson (1964) argued that the results in Figure 7.4.3 are inconsistent. Since the moduli G' and $(G'' - \omega\eta_S)$ are determined by the same spectrum of relaxation times, it is possible to calculate one modulus uniquely from the other in a given polymer–solvent system. Clearly this is not possible for the data of Figure 7.4.3. The experimental results for $(G'' - \omega\eta_S)$ should correspond to values of G' which are also on the Zimm curve.

In order to make the experimental results for the two moduli self-consistent, it is necessary to assume that not all of the polymer contribution to the solution viscosity is able to take part in viscoelastic relaxation. If η_S is replaced by a larger quantity η_∞, G' and $(G'' - \omega\eta_\infty)$ can then be made self-consistent as shown in Figure 7.4.4 for the same polystyrene

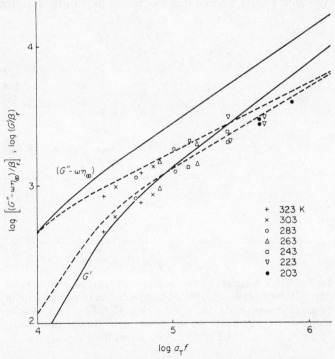

FIGURE 7.4.4. Results of Lamb and Matheson (1964) for G' and $(G'' - \omega\eta_\infty)$ for a 1·15% solution of polystyrene in toluene: $\bar{M}_w = 3\cdot65 \times 10^5$; $\bar{M}_w/\bar{M}_n = 1\cdot08$; results reduced to 303·2 K; $\eta_\infty = \eta_S + 2\cdot0 \times 10^{-4}$ N s m^{-2}; ----, predictions of the Rouse theory replacing η_S by η_∞; —, predictions of the Zimm theory replacing η_S by η_∞.

sample as in Figure 7.4.3. The viscoelastic behaviour of the higher molecular weight samples of polystyrene in toluene is now close to the predictions of the Rouse theory using the correct molecular weight of the polymer. The lower molecular weight polymers could also be described by the Rouse theory with the correct molecular weight and $\eta_\infty > \eta_S$. As the solvent became thermodynamically less good, however, the viscoelastic behaviour deviated towards the non-free-draining Zimm theory.

The existence of a non-relaxing η_∞ greater than η_S can only be inferred for solutions of polystyrene in toluene since current experimental techniques do not permit direct measurement of η_∞ in solvents of low viscosity. In a more viscous solvent, however, where the relaxation times for the motion of polymer segments are longer, it is possible to determine η_∞ experimentally. This has been done for polystyrene in di-n-butyl phthalate and in Arochlor. Figure 7.4.5 contains a combination of the low frequency

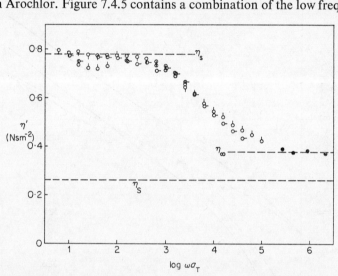

FIGURE 7.4.5. Results of Ferry and coworkers (1966) for the real part of the dynamic viscosity of a 2% solution of polystyrene in pentachlorodiphenyl: $\bar{M}_w = 8.2 \times 10^4$; $\bar{M}_w/\bar{M}_n = 1.05$; results reduced to 298.2 K; $\ominus$, Birnboim and Ferry apparatus, 279.6 K; $\bigcirc$, 283.2 K; φ, 298.2 K; $\bullet$, torsional crystal.

results from the Birnboim and Ferry instrument with some higher frequency results from the torsional crystal for a 2% solution of polystyrene of $\bar{M}_w = 8.2 \times 10^4$ in Arochlor (Ferry and coworkers, 1966). A substantial part of the polymer contribution to the solution viscosity does not show viscoelastic relaxation in this frequency range. This is in agreement with the theory of Peterlin (1967), although the value of η_∞ cannot be calculated from first principles at present.

All of the results discussed so far have been obtained in solutions of finite concentration. The intermolecular entanglements which probably occur in such solutions will affect the dynamics of the polymer molecules, and render suspect any comparison with molecular theories developed for isolated molecules. This problem has been investigated by Sakanishi (1968) using the resonant torsional quartz crystal (Section 6.6) to study dilute polyisobutylene solutions around room temperature. He extrapolated measurements at various low concentrations to infinite dilution, and found that the viscoelastic behaviour changed considerably as the concentration decreased. For polyisobutylene in benzene at the θ-temperature of 297·2 K, there was excellent agreement with the predictions of the unmodified Zimm theory at zero concentration, but deviations occurred at finite concentrations. In the good solvent cyclohexane, the viscoelastic behaviour of polyisobutylene solutions at zero concentration was intermediate between the Rouse and Zimm predictions. Similar behaviour is shown by solutions of poly-α-methyl styrene (Tanaka and Sakanishi, 1969) while the change from Rouse to Zimm behaviour with decreasing concentration has also been observed in very low-frequency measurements on solutions of poly-styrene in pentachlorodiphenyl (Holmes and Ferry, 1968).

Thus the general trend of the available experimental results favours the Zimm description of the viscoelastic behaviour of dilute polymer solutions near the θ-temperature. As the solvent becomes thermodynamically more favourable, the behaviour shifts in the direction of the Rouse free-draining model. This behaviour is complicated by intermolecular entanglements in solutions of finite concentration, and at high frequencies the internal viscosity of the polymer molecule is important. Moreover in small polymer molecules and in larger molecules when the length of the segment becomes so short that its ends no longer obey Gaussian statistics the Rouse and Zimm approaches are not valid (Thurston and Schrag, 1966). All of these effects can be incorporated into the molecular theories (Thurston and Peterlin, 1967), but lack of knowledge of the relative importance of the different effects precludes a detailed quantitative prediction of the viscoelastic behaviour of dilute polymer solutions at present.

The viscoelastic properties of dilute solutions of rigid polymer molecules have been studied less intensively than those of randomly coiled molecules. Figure 7.4.6 shows the results of Tschoegl and Ferry (1964b) for a 1 % solution of poly-γ-benzyl-L-glutamate in the helicogenic solvent m-methoxy phenol, together with the predictions of the Kirkwood—Auer (1951) theory. At low frequencies there is good agreement between experiment and theory, the discrepancy found in the molecular weight of the polymer being probably caused by the molecular weight distribution. The Zimm theory cannot be applied to the results as it would require unrealistically large values of molecular weight in this region. This suggests that the molecular helix is responding to low-frequency deformations as a rigid rod. At higher frequencies, however, there are deviations from the Kirkwood–Auer theory, which indicate some flexibility of the helix: these deviations

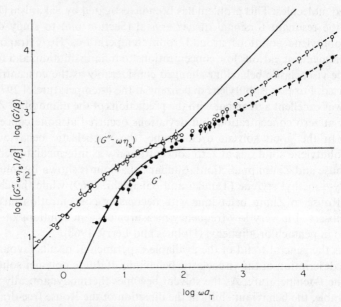

FIGURE 7.4.6. Results of Tschoegl and Ferry (1964b) for G' and $(G'' - \omega\eta_s)$ for a 1% solution of poly-γ-benzyl-L-glutamate in m-methoxy phenol: $\bar{M}_w = 3 \cdot 45 \times 10^5$; results reduced to 298·2 K; $\ominus$, 273·1 K; $\ominus$, 298·2 K; $\ominus$, 323·5 K; —, predictions of the Kirkwood–Auer theory using $\bar{M}_w = 6 \cdot 8 \times 10^5$; ---, 'best line' through experimental points.

are accentuated by increasing the polymer concentration. Similar behaviour has been observed in poly-γ-methyl-D-glutamate (Tanaka and coworkers, 1966) and in shear degraded deoxyribonucleic acid (Meyer, Pfeiffer and Ferry, 1967).

Figure 7.4.7 gives a comparison of the results of Allis and Ferry (1965) for a 0·33% solution of paramyosin in aqueous glycerol with the Cerf (1952) and Scheraga (1955) theory for ellipsoidal molecules. Paramyosin has a stable, helical structure with considerable stiffness, and there is reasonable agreement with the theoretical predictions which assume a single relaxation time for the rotation of the molecule. The increase in G' at high frequencies can again be attributed to a slight bending of the molecule.

Allis and Ferry (1965) have also studied the viscoelastic properties of solutions of the corpuscular protein bovine serum albumin in aqueous glycerol. The native protein exhibits no viscoelastic relaxation in the available frequency range, suggesting that the molecules are behaving as rigid spheres. After denaturation by ageing or by treatment with acid, however, the rigid molecular structure appears to be loosened and the resulting

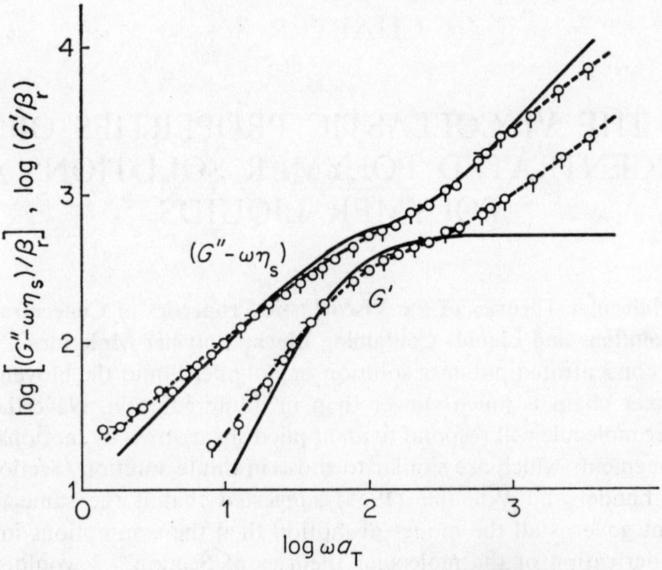

FIGURE 7.4.7. Results of Allis and Ferry (1965) for G' and $(G''-\omega\eta_S)$ of a 1% solution of paramyosin in 80% glycerol–water mixture containing 0·6 M KCl: results reduced to 273·2 K; ♀, 263·2 K; -○, 273·2 K; —, predictions of the Cerf–Scheraga theory, with $\bar{M}_w = 3{\cdot}4 \times 10^5$; ---, 'best line' through experimental points.

viscoelastic behaviour is intermediate between that corresponding to a flexible coil and to a rigid rod. It is evident that viscoelastic studies of such partially flexible molecules in solution can give a sensitive indication of molecular flexibility, and they should be particularly valuable for investigating short-range molecular motions (Dev, Lochhead and North, 1970).

CHAPTER 8

THE VISCOELASTIC PROPERTIES OF CONCENTRATED POLYMER SOLUTIONS AND POLYMER LIQUIDS

8.1 Molecular Theories of the Viscoelastic Properties of Concentrated Solutions and Liquids Containing Linear Polymer Molecules

In a concentrated polymer solution or polymer liquid the movement of a polymer chain is much slower than in dilute solution. Nevertheless a polymer molecule will respond to an applied shear stress by motions of the chain segments which are similar to those in dilute solution (Section 7.1). Ferry, Landel and Williams (1955) suggested that if the same friction constant governs all the modes of motion then the assumptions involved in the derivation of the molecular theories of Section 7.2 would still be valid. Most discussions of the viscoelastic behaviour of polymer liquids have been carried out in terms of the Rouse (1953) theory, although there is no experimental evidence that this should be preferred over the Zimm (1956) theory. Since the solvent viscosity is negligible (or zero) compared to the solution viscosity, equation (7.2.9) can be rewritten in terms of the polymer density ρ to give the relaxation times of the modes of motion of a polymer chain in a concentrated solution or in a polymer liquid:

$$\tau_p = 6\eta_s M / \pi^2 p^2 \rho RT, \quad p = 1, 2, ..., N \qquad (8.1.1)$$

Combination of equation (8.1.1) with equations (7.2.3) and (7.2.4) gives expressions for the frequency dependence of G', G'' and η' of the liquid. The shapes of these curves are the same as those shown in Figure 7.2.1.

It is found experimentally that many concentrated polymer solutions and polymer liquids deviate from this type of behaviour to show a pronounced plateau region in which G' and G'' do not increase with frequency over a certain range of frequencies. This plateau is attributed to the entanglement coupling of the polymer molecules (Ferry, 1969).

An entanglement is a temporary cross-link formed when two polymer molecules become mechanically entwined although no chemical cross-link is present. The entanglement network is able to store elastic energy for short times in the same way as a permanently cross-linked rubber. At longer times the entanglements move, and the polymer is able to flow. Although the most convincing proof of the existence of entanglements is obtained from viscoelasticity studies, supporting evidence is obtained from studies of the steady flow viscosity of polymers. In polymer liquids of moderately low molecular weight, the viscosity increases in direct proportion to the molecular weight of the polymer. At higher molecular weights, however, the viscosity increases with the 3·4 power of the molecular weight. (Figure 8.1.1). Above a critical chain length which is characteristic

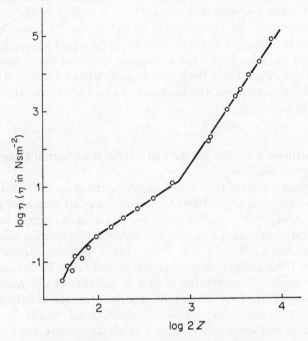

FIGURE 8.1.1. Dependence of steady flow viscosity upon number of chain atoms per molecule for polystyrene fractions at 490 K: Z is the degree of polymerization (after Fox, Gratch and Loshaek, 1956).

of the individual polymer, there are at least two entanglement linkages per molecule so that the motion of all the molecules in the system is effectively coupled. In such an entangled system the effective friction constant for the lower modes of motion which involve the cooperative movement of large portions of the polymer chain is greatly increased. Since these low frequency modes account for the major part of steady flow viscosity of the polymer, Ferry, Landel and Williams (1955) suggested that equation (8.1.1) could still be used to predict the viscoelastic behaviour of polymer liquids and solutions in the terminal zone where the period of an applied shear wave is greater than the mean lifetime of the entanglements.

This approach can be used until the length of a mode of motion of the chain becomes equal to the average distance between entanglements in a molecule. Ferry, Landel and Williams (1955) argued that the motion of higher modes will be unaffected by entanglements and that this motion will be governed by the monomeric friction coefficient ζ_0 ($\zeta_0 < \zeta$). Barlow, Harrison and Lamb (1964) have suggested that ζ_0 may be calculated from the steady flow viscosity which would have existed in the absence of entanglements, η_x, by extrapolating at unit slope the lower molecular weight line in Figure 8.1.1. This value of η_x is then used in equation (8.1.1) to calculate the relaxation times at frequencies above the plateau region. Thus all relaxation times in the higher frequency block are shifted to higher frequencies in comparison with the block of relaxation times in the terminal zone by the ratio ζ/ζ_0.

8.2 Experimental Results for the Viscoelastic Behaviour of Concentrated Polymer Solutions

The earliest studies of the viscoelastic properties of concentrated polymer solutions were generally confined to frequencies of alternative strain below 1000 Hz. Using the wave propagation (Section 6.2) and transducer (Section 6.4) methods, Ferry and his collaborators studied solutions of polyisobutylene (Ashworth and Ferry, 1949), polyvinylacetate (Sawyer and Ferry, 1950) and polystyrene (Grandine and Ferry, 1953) over a range of temperature and concentration and in solvents of different thermodynamic character. They devised an empirical reduced variables method which allowed the superposition of experimental results at different temperatures and concentrations on a single universal curve for a given polymer–solvent system. This principle of reduced variables was modified and given a firm theoretical basis with the development of the Rouse theory in 1953, and Philippoff (1954) showed that equations (7.3.3), (7.3.4) and (7.3.5) with $\eta_S = 0$ may be applied exactly to concentrated polymer solutions.

The comparison of the experimental results with the predictions of the Rouse theory was first undertaken for concentrated polymer solutions by Ferry, Williams and Stern (1954). By assuming that the relaxation times of the first few modes of motion of the polymer chain were all prolonged by the same amount by molecular entanglements, they were able to obtain moderate agreement between experiment and theory for a number of vinyl polymer fractions in the terminal zone of viscoelastic relaxation. The agreement was least satisfactory in those polymers which had a broad molecular weight distribution, the discrepancies being particularly marked in a blend of high and low molecular weight fractions. This simple modification of the Rouse theory is also unable to account for the entanglement plateau which is observed in solutions of higher molecular weight polymers: the further modification of the Rouse theory described in Section 8.1 would probably describe this behaviour satisfactorily, although this latter approach has hitherto been applied only to polymer liquids.

An interesting comparison of the viscoelastic properties of solutions of cellulose derivatives with those of vinyl polymers has been made by Landel and Ferry (1955) and Plazek and Ferry (1956). Although the viscoelastic behaviour of a number of vinyl polymers of the same molecular weight is similar in the plateau region, cellulose derivatives have a longer and flatter plateau. In particular, cellulose tributyrate and cellulose triacetate solutions have a spectrum of relaxation times which is perfectly flat over four decades of frequency. As discussed in Section 8.1, the length of the plateau region is determined by the ratio of the entangled to the monomeric friction constants, ζ/ζ_0. Cellulose derivatives have a rigid chain backbone and a cellulose molecule of a given molecular weight is much more extended than a vinyl molecule of the same molecular weight. Hence cellulose derivatives will be more entangled than the corresponding vinyl polymers, and the increased value of the entangled friction constant ζ will lead to a more extensive plateau region.

Although the concept of molecular entanglement is qualitatively satisfactory for explaining the plateau region, certain discrepancies occur when the modified Rouse theory is applied quantitatively. It would be expected that the average number of chain bonds between entanglement points would be proportional to the volume fraction of polymer in solution if there is a constant probability of entanglement of two adjacent monomer units belonging to neighbouring chains. Such behaviour has been found for concentrated solutions of polyisobutylene in n-hexadecane (Richards, Ninomiya and Ferry, 1963), but for a number of solutions of alkyl methacrylates the entanglement coupling density shows an anomalous dependence on temperature and concentration: for example, entanglement points are required to dissociate at high temperatures (Berge, Saunders and Ferry, 1959) and the average degree of polymerization between entanglement points to increase as the 2·3 power of polymer concentration (Saunders and coworkers, 1959). Bueche, Coven and Kinzig (1963) have pointed out that these difficulties arise from the arbitrary way in which the dilute solution theory of Rouse has been modified, first for concentrated solutions and then for entanglements: for example, the assumption that the nature and length of the freely orienting segment do not change with concentration and temperature is unlikely to be strictly true. Although the modified Rouse theory gives good qualitative agreement with experiment, a new theoretical treatment is required to give a precise quantitative description of concentrated polymer solutions of linear molecules.

Moore and coworkers (1967a) have investigated the limiting high frequency viscosity η_∞ (Section 7.4) of concentrated polymer solutions. For solutions of monodisperse polystyrene in di-n-butyl phthalate, they found a finite value of η_∞ at low concentrations. But as the concentration of the polymer increases and entanglement sets in, η_∞ becomes negligible. Moore and coworkers (1967a, b) also attempted to test the relative merits of the Rouse and Zimm theories for concentrated solutions. At high concentrations there is good agreement with the Rouse theory, although as the concentration decreased there was a deviation in the direction of the Zimm theory. Unfortunately the experimental accuracy was insufficient to define η_∞ accurately or to establish exactly the role of hydrodynamic interaction in concentrated solutions.

8.3 Experimental Results for the Viscoelastic Behaviour of Polymer Liquids

Although the viscoelastic behaviour of polymer liquids has been widely studied, many of the results are not capable of quantitative interpretation. Much work, in particular that on polyethylene and polypropylene, has been carried out on polymer liquids below their melting-points: the viscoelastic behaviour of such partly crystalline substances is not reproducible from one sample to another even though X-ray studies show that only a small amount of crystallinity is present. Further complications arise from chain branching and from anomalous molecular weight distributions in certain polymers.

The first study of the viscoelastic properties of polymer liquids was that of Philippoff (1934) who investigated cellulose acetate: he found an increase in the rigidity modulus and a decrease in the dynamic viscosity with increasing frequency of alternating strain. Subsequently Tobolsky and McLoughlin (1952) and Fitzgerald, Grandine and Ferry (1953) demonstrated the existence of a plateau in the relaxation spectrum of high molecular weight polyisobutylenes. After Ferry, Landel and Williams (1955) had modified the Rouse theory to take account of molecular entanglements, Cox, Nielsen and Keeney (1957) showed that a number of polystyrene fractions gave qualitative agreement with the modified Rouse theory.

One of the most successful investigations of the viscoelastic behaviour of polymer liquids has been that of Barlow, Harrison and Lamb (1964) and Lamb and Lindon (1967): they studied a series of polydimethylsiloxane liquids having a 'most-probable' distribution of molecular weights over a range of frequencies from 20 Hz–78 MHz and from 220–320 K.

It was found possible to reduce the experimental data for the components of the complex rigidity modulus of each liquid to a single curve by the method of reduced variables. Figure 8.3.1 shows a comparison of the experimental results for a polydimethylsiloxane sample of $3 \cdot 5 \times 10^{-4}$ m^2 s^{-1} viscosity and $\bar{M}_n = 1 \cdot 58 \times 10^4$ with the predictions of the Rouse theory for

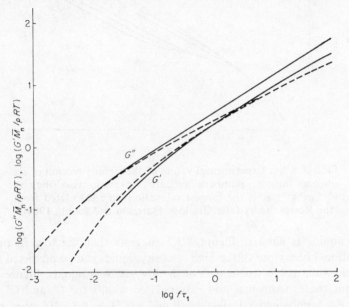

FIGURE 8.3.1. Comparison of experimental results (—) with predictions of Rouse theory (---) for the components of the rigidity modulus of a silicone liquid of kinematic viscosity 350×10^{-6} m^2 s^{-1}: τ_1 is the longest relaxation time calculated from the Rouse theory (after Barlow, Harrison and Lamb, 1964).

an undiluted polymer of most probable molecular weight distribution. At low frequencies there is good agreement between theory and experiment. The deviations at higher frequencies probably occur because the length of the chain involved in the higher modes of motion is less than the minimum chain length required to give the assumed Gaussian distribution of end-to-end distances of the segment. This occurs when the average number of chain bonds in the segment is 20, a figure which is smaller than that for vinyl polymers but which is reasonable for such a flexible polymer chain.

For polydimethylsiloxanes of higher molecular weight, a plateau region is observed. There is good agreement with the predictions of the Rouse theory in the terminal zone. Figure 8.3.2 shows the normalized rigidity modulus G' of six liquids plotted against $\log f\tau_1$, where τ_1 is the relaxation time of the longest mode of motion of the polymer chain calculated from the Rouse theory by equation (7.2.9) with $\eta_S = 0$; the behaviour of G'' of

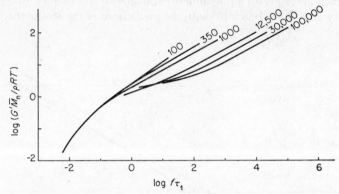

FIGURE 8.3.2. Experimental values of the rigidity moduli of six silicone liquids; numbers indicate kinematic viscosities in 10^{-6} m^2 s^{-1}: τ_1 is the longest relaxation time calculated from the Rouse theory (after Barlow, Harrison and Lamb, 1964).

these liquids is similar. Figure 8.3.2 suggests that the low frequency relaxational behaviour of the high viscosity liquids can be predicted from the behaviour at low frequencies of the low viscosity liquids. Figure 8.3.3 contains the experimental and extrapolated results for G' and G'' of a 10^{-1} m^2 s^{-1} polydimethyl siloxane liquid of $\bar{M}_n = 6\cdot8 \times 10^4$ over seven decades of frequency. The theoretical curves above the plateau have been obtained using a monomeric friction coefficient calculated as described in Section 8.1. There is excellent agreement between theory and experiment over the entire frequency range.

It is therefore possible to predict the viscoelastic behaviour of the polydimethylsiloxane liquids from a knowledge of the steady flow viscosity and molecular weight: only when the length of the polymer segment is less than 20 chain bonds is the theory inadequate. Alternatively, if the viscosity of the polymer as a function of molecular weight is not known, the average number of chain bonds between entanglement points Λ may be calculated from the onset of the plateau region of viscoelastic behaviour. Although few data are available, Markovitz, Fox and Ferry (1962) have shown that Λ values calculated from viscosity and from viscoelasticity are in reasonable agreement.

It should be emphasized, however, that the Rouse theory was developed for polymer solutions at infinite dilution, and that subsequent modifications of this theory for polymer liquids are wholly empirical. The assumption underlying these modifications is that entanglements affect all the longest relaxation times in a polymer liquid to the same degree. Some measurements of Sanders, Ferry and Valentine (1968) suggest that this approximation may have its limitations.

The transition from polymers to simple liquids has been investigated by Barlow, Dickie and Lamb (1967). They found that a sample of poly-1-butene with $\bar{M}_n = 448$ behaves as a simple liquid. Samples of $\bar{M}_n = 640$

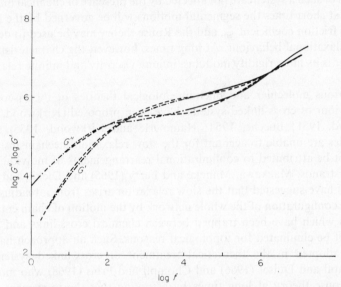

FIGURE 8.3.3. Comparison of experimental results (—) with the predictions of the Rouse theory modified to allow for entanglements (---) for the components of the rigidity modulus of a silicone liquid of kinematic viscosity 10^{-1} m^2 s^{-1}: (-·-·) represents extrapolated experimental results (after Barlow, Harrison and Lamb, 1964).

and above, however, have an additional low frequency relaxation which was attributed to 'quasi-Rouse' modes of motion of the polymer chain. These are analogous to the polymer modes discussed in Section 7.1, although the polymer chain is not sufficiently long for the distribution of chain ends to obey Gaussian statistics. These results could not be interpreted unambiguously as the samples of poly-1-butene were polydisperse and highly branched. Similar results were obtained from a series of poly-ethyl and poly-n-butyl acrylates. Once again the short-time behaviour was similar to that of a simple liquid while the long-time relaxations could be described by a number of Rouse modes (Barlow and coworkers, 1969). Nevertheless the region between a simple liquid and a Gaussian chain remains ill-defined, and further experimental investigation would be desirable.

8.4 The Viscoelastic Properties of Cross-linked Polymers

A permanently cross-linked rubber may be considered to be composed of one giant molecule in which the individual polymer chains are joined together by chemical bonds. The relatively short-range segmental rearrangements of such a system are not affected by the presence of chemical linkages. Thus at short times the segmental motions will be governed by the monomeric friction coefficient ζ_0, and the Rouse theory may be used to describe the relaxational behaviour. At long times, however, the characteristic of a rubber is its finite rigidity modulus, infinite viscosity and infinite relaxation time.

Various molecular and phenomenological theories of the viscoelastic behaviour of cross-linked systems have been proposed (Kirkwood, 1946; Blizard, 1951; Bueche, 1954; Hammerle and Kirkwood, 1955). These theories are unable to account for the slow relaxation mechanisms which cannot be attributed to configurational rearrangements of individual network strands. Maekawa, Mancke and Ferry (1965) and Dickie and Ferry (1966) have suggested that the slow relaxation arises from a readjustment of the configuration of the whole network by the motion of chain entanglements which have been trapped between chemical cross-links and which cannot be eliminated for topological reasons. Such an approach has also been adopted in the theoretical treatment of cross-linked systems by Chompff and Duiser (1966) and Chompff and Prins (1968) who modified the Rouse theory at long times by assuming that the movement of an entanglement along a polymer chain involves a frictional resistance.

For complete trapping of an entanglement, each of the four strands radiating from it must be permanently attached to the network. If one strand corresponding to the end of a polymer chain is unattached, slow uncoiling of the entanglement is possible. In lightly cross-linked rubbers this gives rise to a very long-time relaxation process (Mancke, Dickie and Ferry, 1968).

8.5 The Viscoelastic Properties of Polymer Glasses

As a polymer liquid is cooled, the viscosity increases and the relaxation times for chain motion increase according to equation (8.1.1). When the longest segmental relaxation time becomes comparable to the duration of a typical experiment (say 10^3 s), then cooperative motion of the chain segments is unable to occur and the liquid does not flow during the time of the experiment. The temperature at which this occurs is called the glass transition temperature, T_g, and it corresponds to a polymer viscosity in the region of 10^{12} N s m^{-2}. At temperatures below T_g the polymer glass is no longer in equilibrium, but the rate of approach to equilibrium is so slow that it can generally be neglected. The free volume which is available at T_g (typically about 2% of the specific volume) is thus apparently frozen in at temperatures below T_g, and is available to permit the movement of small chain segments and of side-chains. As the temperature of the glass is lowered, the decreased thermal energy of the moving segments slows down their movement until it becomes comparable to the time scale of the observation and a subsidiary T_g is observed. It is important to note that

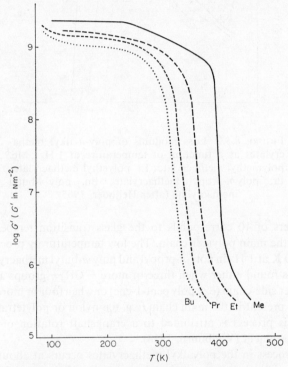

FIGURE 8.5.1. Rigidity modulus of poly-n-alkyl methacrylates as a function of temperature at 1 Hz: Me, polymethyl methacrylate; Et, polyethyl methacrylate; Pr, poly-n-propyl methacrylate; Bu, poly-n-butyl methacrylate (after Heijboer, 1965).

both the main and subsidiary T_g's arise from relaxation processes, and the values observed vary with the time scale of the experiment.

Several different relaxation processes are observed in most glassy polymers. The most widely studied series is the polyalkyl methacrylates. Figures 8.5.1 and 8.5.2 show the variation of the rigidity and loss moduli of four methacrylate polymers as a function of temperature at a frequency of alternating shear strain of 1 Hz (Heijboer, 1965). Three loss maxima are observed, and these are commonly called the α, β and γ processes. The high temperature α process in which the rigidity modulus falls by over

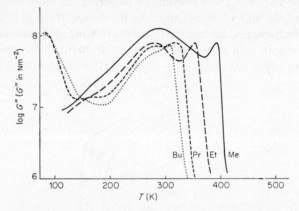

FIGURE 8.5.2. Loss modulus of poly-n-alkyl methacrylates as a function of temperature at 1 Hz: Me, polymethyl methacrylate; Et, polyethyl methacrylate; Pr, poly-n-propyl methacrylate; Bu, poly-n-butyl methacrylate (after Heijboer, 1965).

three powers of 10 corresponds to the glass transition temperature for motion of the main polymer chain. The low temperature γ process occurs at about 90 K at 1 Hz in poly-n-propyl and poly-n-butyl methacrylate. Such a process is found either when three or more $-CH_2-$ groups are present in a polymer side-chain (e.g. poly pent-1-ene) or when four or more $-CH_2-$ groups are present in the main chain (e.g. 6,6-nylon or polytetramethylene oxide). This process is attributed to a crankshaft rotation of the poly-methylene segment (Willbourn, 1958).

The β process in the polyalkyl methacrylates occurs at about 290 K at 1 Hz. Although many investigations have been carried out on this process, its origin is not completely clear. It is likely that some form of motion of the ester group about the C—C bond which links it to the main chain is involved. This motion is considerably hindered by the α-methyl group in the main methacrylate chain, and when it is absent, as in polymethyl acrylate, the β maximum occurs at a substantially lower temperature in a constant frequency experiment. Other molecular motions have been detected in the methacrylates by n.m.r.: in polymethyl methacrylate and polyethyl methacrylate, for example, motion of the ester methyl and ethyl groups occurs down to temperatures below 77 K, whereas the more sterically hindered α-methyl group does not begin to rotate until the temperature exceeds 130 K (Sinnott, 1960).

A further relaxation process is observed in glassy polycycloalkyl methacrylates. In polycyclohexyl methacrylate, for example, the process occurs at 190 K for 1 Hz and has an activation energy of 48 kJ mol^{-1}. This is caused by the chair–chair inversion of the cyclohexyl group. Similar behaviour is shown by polycyclooctyl methacrylate, while the more flexible rings of polycyclopentyl and polycycloheptyl methacrylates show relaxation at lower temperatures (Heijboer, 1965, 1968).

The relaxational behaviour of glassy polystyrene is similar to that of the acrylate polymers. A β transition is found close to T_g and is attributed either to a phenyl group motion (Illers, 1961) or to a local oscillation of the main chain (Saito and coworkers, 1963). The γ process is generally small and appears to be caused by structural irregularities in the polymer such as tail-to-tail addition of monomers during polymerization (Illers and Jenckel, 1959). A low temperature δ process has also been found in *para*-substituted polystyrenes although not in the *ortho*- or *meta*-substituted polymer (Baccaredda and coworkers, 1966).

A number of other glassy polymers have been investigated. Although there is often a satisfactory qualitative explanation of the observed relaxation processes, many anomalies remain. These probably arise from the attempt to treat the process under investigation as an isolated relaxation despite the fact that the motions of side-groups and chain segments are strongly coupled. Such problems could be considerably clarified if a range of relaxation techniques (mechanical, ultrasonic, dielectric and n.m.r.) were applied to a number of closely related polymer systems.

CHAPTER 9

THE VISCOELASTIC PROPERTIES OF SIMPLE LIQUIDS

9.1 Introduction

In this chapter 'simple' liquids are taken to be non-polymeric liquids, that is atomic liquids or liquids in which the constituent molecules have less than about 20 chemical bonds in their chain backbones. To allow quantitative theoretical treatment of the viscoelastic behaviour of liquids, it would be desirable to study experimentally liquids having the simplest possible structure such as liquid argon or methane. The viscosity of such liquids is typically 10^{-3} N s m^{-2}, while G_∞ is likely to be 10^8 N m^{-2} or above. The Maxwell equation (5.4.6) suggests that the viscoelastic relaxation time τ_s would be less than 10^{-11} s in such liquids so that frequencies of alternating strain greater than $(2\pi\tau_s)^{-1} \approx 10^{10}$ Hz would be required to define the viscoelastic relaxation region adequately. Although shear waves in this frequency range have been used to study solids, investigations of the viscoelastic behaviour of non-polymeric liquids have been confined to lower frequencies and to liquids of viscosity greater than 10^{-2} N s m^{-2} (Chapter 6). High viscosities are achieved by cooling a liquid: unfortunately most liquids crystallize a few degrees below their normal melting points where the viscosity is still low. Only liquids of complex molecular structure can be supercooled far below their melting points to give high viscosity glasses, and viscoelastic studies have been restricted mainly to such liquids.

It is found experimentally that the viscoelastic relaxation of such liquids cannot be described by a single Maxwell relaxation time (Section 5.4) and that a distribution of relaxation times (Section 5.5) extending typically over three decades of time is observed. Hence in order to define viscoelastic relaxation completely in a liquid at a single temperature, the measuring system would require to cover at least three decades of frequency. Most techniques for simple liquids operate either at a single frequency or at a small number of fixed frequencies within a range of about one decade. Thus it is customary for viscoelastic relaxation to be studied over a range of temperature as well as of frequency, and for the results to be reduced to some reference temperature by the principle of reduced variables as described in Section 5.7. Such a reduction requires that the steady flow viscosity and G_∞ of the liquid be known throughout the temperature range studied.

Steady flow viscosities may be measured up to 10^2 N s m^{-2} using capillary viscometers and at higher viscosities with a cone and plate or concentric cylinder viscometer. As systems operating above 10^2 N s m^{-2} cannot readily be adapted to the low temperatures where many visco-elastic studies of simple liquids must be conducted, viscosity values of less than 10^2 N s m^{-2} are often extrapolated into the higher viscosity region using an assumed viscosity–temperature relation.

Many equations have been proposed to describe the temperature dependence of liquid viscosities, and two equations have found particularly wide application. The first of these is the Arrhenius or Andrade (1934) viscosity equation which in the integrated form may be written

$$\ln \eta_s = A + E/RT \qquad (9.1.1)$$

A is a constant, R is the gas constant and E is an activation energy for viscous flow. An equation of similar form to this has been derived by Eyring (1936) and developed by Glasstone, Laidler and Eyring (1941) from the hole theory of liquids in which the energy E is expressed as the sum of the energy required to create a hole and of the activation energy required for a molecule to flow into a hole which is already present. Although the Arrhenius equation accurately describes the temperature dependence of the viscosities of many liquids of low viscosity well above their melting points, it often fails for liquids near their melting points and is not applic-able to the supercooled or associated liquids which are frequently used in viscoelastic studies.

At temperatures where the Arrhenius equation fails, the empirical equation proposed by Vogel (1921), Fulcher (1925) and Tammann and Hesse (1926) is often used:

$$\ln \eta_s = A' + B'/(T - T_0) \qquad (9.1.2)$$

A', B' and T_0 are constants, and the liquid viscosity is infinite at the temperature T_0. In most liquids density is linear with temperature to a good approximation and Barlow, Lamb and Matheson (1966) showed that equation (9.1.2) is identical with the empirical Doolittle (1951, 1957) free volume equation

$$\ln \eta_s = A'' + B'' v_0/v_f \qquad (9.1.3)$$

if v_0 is taken as the specific volume of the liquid at T_0. A'' and B'' are constants, while the free volume in the liquid v_f is defined as

$$v_f = v - v_0 \qquad (9.1.4)$$

where v is the specific volume of the liquid at the temperature at which the viscosity is measured. At the temperature T_0, the free volume in the liquid is zero. Equations similar to (9.1.3) have been derived theoretically by Cohen and Turnbull (1959) and Kumar (1963). In some liquids equation (9.1.2) adequately describes the temperature dependence of viscosity in the non-Arrhenius region. In other liquids, however, it is necessary to apply equation (9.1.2) with two different sets of values of the constants A', B' and T_0: in every case the T_0 of the higher temperature non-Arrhenius viscosity region is higher than the T_0 of the lower temperature form (Barlow, Lamb and Matheson, 1966).

This separation of the temperature dependence of viscosity into an Arrhenius and two non-Arrhenius regions means that no less than eight constants are required to describe liquid viscosity from the boiling point ($\sim 10^{-3}$ N s m^{-2}) to the glass transition temperature ($\sim 10^{12}$ N s m^{-2}). Davies and Matheson (1966, 1967) have suggested that the three viscosity regions arise because of the rotational behaviour of the liquid molecules. In the Arrhenius region molecules are able to rotate many times about at least two molecular axes during the time between molecular translational jumps. In the higher temperature non-Arrhenius region molecules can rotate about only one axis during this time, while in the lower temperature non-Arrhenius region molecular rotation occurs primarily as a result of molecular translational motion.

The changeover in viscous behaviour in the non-Arrhenius region typically occurs at a viscosity of about 10^{-1} N s m^{-2}. If viscosity data from U-tube viscometers are available from this viscosity to 10^2 N s m^{-2}, the lower temperature form of equation (9.1.2) can be used to interpolate and extrapolate the liquid viscosity when this is necessary in applying the principle of reduced variables to the viscoelastic experimental data. The same equation can be used to predict the glass transition temperature T_g of the liquid. T_g is the temperature at which the relaxation time for molecular translational motion becomes equal to the duration of an experiment to measure properties of the liquid such as its heat capacity or density (Kauzmann, 1948). For a normal laboratory experiment of 10^3 s, the Maxwell equation (5.4.6) suggests that T_g will correspond to a liquid viscosity of 10^{12} N s m^{-2}. The temperature at which this occurs may be calculated from the lower temperature form of equation (9.1.2): the value of T_g calculated from viscosity data in this way is in good agreement with the value of T_g measured directly by differential thermal analysis (Carpenter, Davies and Matheson, 1967). At temperatures below the glass transition temperature the liquid is a glass: molecular translational motion is not possible during a normal experiment of 10^3 s duration and the liquid is unable to contract to its equilibrium specific volume. The expansion coefficient and heat capacity have values close to those expected for a crystalline solid. It must be stressed that the temperature at which T_g is observed depends on the time-scale of the experiment (Kovacs, 1958). At the short times of observation associated with high frequency shear waves, the apparent T_g occurs in the viscoelastic relaxation region at a temperature well above that of the conventional T_g. At infinitely long times of observation, on the other hand, T_g would coincide with T_0.

Although equation (9.1.2) is a useful approximation to describe the temperature dependence of the viscosity of liquids over a moderate range of temperature, the accuracy of extrapolated values of the viscosity of a given liquid is unknown. The difficulties of devising a satisfactory viscosity–temperature relation have been described by Goldstein (1969). In applying the method of reduced variables to extend the effective frequency of measurement of the viscoelastic properties of liquids, extrapolated viscosity values should be treated with considerable caution.

9.2   Experimental Results for G_∞ of Liquids

At sufficiently high frequencies of alternating strain a liquid behaves purely elastically, its rigidity modulus G_∞ being comparable to that of a solid and its dynamic viscosity being zero. Figure 9.2.1 shows a typical set of experimental results obtained by Barlow and coworkers (1966) for

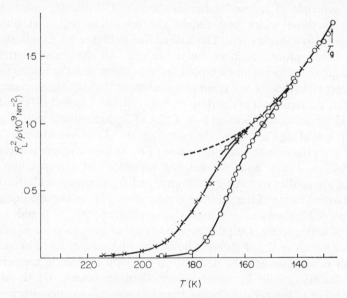

FIGURE 9.2.1. R_L^2/ρ for *sec*-butyl benzene as a function of temperature: ○, 30 MHz; ×, 450 MHz; □, 1000 MHz; T_g, measured glass transition temperature; ---, values of G_∞ calculated by method of reduced variables (after Barlow and coworkers, 1966).

sec-butyl benzene using shear waves of frequencies 30, 450 and 1000 MHz over a range of temperature. At low temperatures where the liquid behaves elastically, X_L is small and $R_L^2/\rho \approx G' = G_\infty$: R_L^2/ρ is then independent of the frequency of measurement at a given temperature.

At higher temperatures the decrease in R_L^2/ρ with increasing temperature is the result not only of a decrease in G_∞ of the liquid with increasing temperature but also of viscoelastic relaxation. When shear waves of only one frequency are used, it is not possible to separate the G_∞ region from the relaxation region. With two or more measuring frequencies, however, the G_∞ region may be identified from the independence of R_L^2/ρ on frequency. The range of temperature over which G_∞ can be determined with shear waves of frequencies below 1000 MHz is comparatively small. The lower temperature limit is the glass transition temperature T_g of the liquid: below T_g the liquid does not attain its equilibrium specific volume within a reasonable time so that measurements can only be made on a non-equilibrium glass. The higher temperature limit is determined by the viscoelastic relaxation region which at a measuring frequency of 1000 MHz occurs at liquid viscosities below 10^3 N s m^{-2}. Some extension of the 'measured' G_∞ range to higher temperatures is attainable by the application of the principle of reduced variables (Section 5.7), the G_∞ values being chosen as those which best reduce the relaxation region at different frequencies to a single curve. The dashed line in Figure 9.2.1. indicates the result of this procedure in *sec*-butyl benzene. An alternative method of estimating G_∞ is to use an extrapolation procedure with the results for R_L at different frequencies at a given temperature in the viscoelastic relaxation region. If the liquid obeys the Barlow, Erginsav and Lamb (1967) model of viscoelastic relaxation (Section 9.4), a plot of R_L against $\omega^{-\frac{1}{2}}$ should give a straight line at high ω values: the extrapolation of this line to a zero value of $\omega^{-\frac{1}{2}}$ gives the infinite frequency value of R_L at that temperature (Hutton and Phillips, 1969). Both the reduced variables and extrapolation procedures are subject to some uncertainty, and G_∞ cannot usually be determined over a range of temperature greater than 30 K in organic liquids.

Figure 9.2.2 shows some representative results for G_∞ of liquids over a range of temperature. Despite the wide range of molecular structure, the absolute values of G_∞ of different liquids lie within an order of magnitude of each other although the decrease in G_∞ with increasing temperature occurs more rapidly in some liquids than in others. G_∞ is almost independent of temperature below the glass transition temperature of the liquids.

Many attempts have been made to describe the temperature dependence of G_∞ mathematically. Meister and coworkers (1960) and Slie, Donfor and Litovitz (1966) find for several associated liquids that G_∞ decreases linearly with increasing temperature over a small temperature range, while Gruber and Litovitz (1964) assume an exponential decrease for $ZnCl_2$. In contrast Barlow and coworkers (1966) found that $1/G_\infty$ increased linearly with temperature for a number of liquids.

Most of the results available for G_∞ of simple liquids cover such a small range of temperature that a number of different descriptions of the results

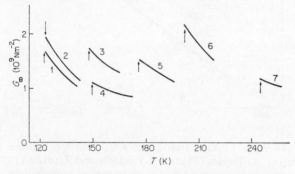

FIGURE 9.2.2. Observed G_∞ values of some liquids: 1, n-propyl benzene; 2, sec-butyl cyclohexane; 3, 3-phenyl propyl chloride; 4, hepta-methyl nonane; 5, 3-phenyl propanol; 6, tri(m-tolyl)phosphate; 7, m-bis(m-phenoxy phenoxy)benzene; arrows indicate measured glass transition temperatures. Results of Barlow, Erginsav and Lamb (1967), Davies (1968) and Hutton and Phillips (1969).

may be made within experimental error. The most pertinent results for establishing the form of the temperature dependence are those of Barlow and coworkers (1966, 1967) for sec-butyl benzene and di(iso-butyl) phthalate, of Capps and coworkers (1966) and Tauke, Litovitz and Macedo (1968) for B_2O_3, of Kono, Litovitz and McDuffie (1966) for n-propanol, and of Piccirelli and Litovitz (1957) and Davies (1968) for glycerol. In each case the temperature dependence of G_∞ is represented within experimental error by the relation (Barlow and coworkers, 1966)

$$1/G_\infty = 1/(G_\infty)_0 + B(T - T_0) \qquad (9.2.1)$$

where the constant $(G_\infty)_0$ may be considered to be the value of G_∞ at the temperature T_0 and B is a constant. Figures 9.2.3 and 9.2.4 show the application of (9.2.1) to glycerol and B_2O_3. The results for a large number of other non-polymeric liquids which have been studied over a smaller range of temperature are consistent with equation (9.2.1), and a few

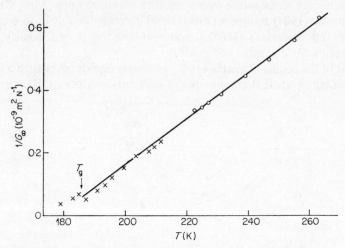

FIGURE 9.2.3. Linear variation of $1/G_\infty$ with temperature for glycerol: ×, Davies (1968); ○, Piccirelli and Litovitz (1957); arrow indicates measured glass transition temperature.

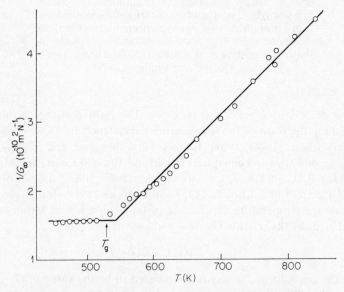

FIGURE 9.2.4. Linear variation of $1/G_\infty$ with temperature for boron trioxide from the results of Capps and coworkers (1966); arrow indicates measured glass transition temperature.

representative values of the constants $(G_\infty)_0$ and B are given in Table 9.2.1. This equation is applicable to a wide range of liquids, including fused salts, associated liquids and hydrocarbons. It is noteworthy, however, that all

TABLE 9.2.1. Parameters of the equation $1/G_\infty = 1/(G_\infty)_0 + B(T - T_0)$

Liquid	$(G_\infty)_0$ (10^9 N m^{-2})	B $(10^{-11} \text{ m}^2 \text{ N}^{-1} \text{ K}^{-1})$	T_0 (K)	Reference*
n-Propyl benzene	2·3	1·5	109·7	1
sec-Butyl benzene	5·6	1·4	98·6	2
sec-Butyl cyclohexane	3·5	1·8	111·0	1
Squalane	3·7	1·5	133·9	3
2-Phenyl ethyl chloride	2·6	1·1	130·7	1
3-Phenyl propyl chloride	6·0	1·4	120·9	1
3-Phenyl propanol	4·0	1·1	137·3	1
1,3-Butanediol	3·6	0·30	120·2	3
Glycerol	13·2	0·18	118·3	1, 4
Di(n-butyl)phthalate	4·8	1·9	151·3	2
Di(iso-butyl)phthalate	6·7	1·3	144·9	2
Tri(o-tolyl)phosphate	1·5	0·40	199·1	3
Tri(m-tolyl)phosphate	7·4	1·3	178·5	3

* References
1. Davies (1968).   3. Barlow, Erginsav and Lamb (1969).
2. Barlow and coworkers (1967). 4. Piccirelli and Litovitz (1957).

these liquids have complex molecular structures: no experimental information is yet available on G_∞ of the simplest atomic and diatomic liquids.

9.3 Theories of G_∞ of Liquids

Many attempts have been made to calculate the absolute value and the temperature dependence of G_∞ of liquids. A statistical mechanical treatment of Green (1952) has been adapted by Zwanzig and Mountain (1965) to give

$$G_\infty = \rho kT + \frac{2\pi}{15} \rho^2 \int_0^\infty g(r) \frac{d}{dr} \left(r^4 \frac{d\phi}{dr} \right) dr \qquad (9.3.1)$$

Here ρ is the liquid density, $g(r)$ is the radial distribution function and ϕ is the intermolecular potential. This expression may be simplified if the intermolecular interactions in the liquid can be represented by a two-body, central intermolecular potential such as the Lennard–Jones potential:

$$G_\infty = \tfrac{26}{5} \rho kT + 3P + \tfrac{24}{5} \rho E \qquad (9.3.2)$$

P is the external pressure acting on the liquid, and E is the internal energy.

A similar approach to that of Zwanzig and Mountain (1965) has been adopted by Dexter and Matheson (1970). They used the liquid model of Bernal (1964) and Scott (1962) which gives the coordinates of a heap of 1005 atoms, the radial distribution function of which resembles that of a monatomic liquid at its triple point. These atoms were assumed to interact with a Lennard–Jones pair potential, and the potential energy of the system U_{LJ} calculated for a number of different deformations. G_∞ of the array is then given by

$$G_\infty = \frac{1}{v} \frac{\partial^2 U_{LJ}}{\partial (\tan \theta)^2} \qquad (9.3.3)$$

where v is the volume of liquid considered and θ is the angle of shear. Variations in density and temperature were simulated by the random removal of atoms from the array, and the results are given in Table 9.3.1.

TABLE 9.3.1. G_∞ of liquid argon under its saturated vapour pressure (Dexter and Matheson, 1970)

T (K)	83·8	88·3	92·9	97·4	101·8
Relative density	1·00	0·98	0·96	0·94	0·92
G_∞ (10^9 N m^{-2})	1·23	1·16	1·03	0·84	0·76

The temperature dependence of G_∞ is similar to that found experimentally in organic liquids (Section 9.2). It is much greater than that predicted from theories based on the inclusion of small holes in a continuum (Dewey, 1947; Phillips, 1969). The results do not provide strong evidence to support a linear temperature dependence of either G_∞ or J_∞. There is good agreement with the molecular dynamics calculations of Verlet (1967), but no experimental results are available for comparison.

In liquids containing more complex molecules, it is not possible to assume such a simple intermolecular potential nor to specify the molecular coordinates. Moreover an important effect of shearing a glass containing flexible molecules will be the changes in molecular conformation. Hence treatments such as that of Zwanzig and Mountain (1965) are restricted to the simplest liquids such as argon or methane. A more promising approach to molecular liquids is that of Dienes (1958) who relates the elastic constants of a liquid to its spectrum of Debye frequencies. The absence of information on the change of the frequencies of the Debye modes with deformation precludes the quantitative application of this theory at present.

There have been many other attempts to calculate G_∞ of liquids and its dependence upon temperature. Eyring and his collaborators have derived expressions for the viscosity of liquids from rate process theory and the significant structure theory: by combining these expressions with the Maxwell relation (5.4.6), G_∞ may be calculated. The rate process expression derived by Hirai and Eyring (1958) is

$$G_\infty = \frac{kT}{v^*} \exp\left(\frac{E_h}{kT}\right) \qquad (9.3.4)$$

where v^* is the volume of a molecule and E_h is the energy necessary to create a hole of volume v^* in the liquid. The alternative model of Ree, Ree and Eyring (1962) assumes that a liquid consists of molecules displaying either gas-like or solid-like properties. In viscous liquids the expression for the viscosity becomes

$$\eta_s = \frac{6RT\tau_s}{2^{\frac{1}{2}} z_n(V - V_S)} \qquad (9.3.5)$$

whence

$$G_\infty = \frac{6RT}{2^{\frac{1}{2}} z_n(V - V_S)} \qquad (9.3.6)$$

τ_s is the viscoelastic relaxation time, z_n is the number of nearest neighbours of a liquid molecule, and V and V_S are the molar volumes of liquid and solid respectively. Equations (9.3.4) and (9.3.6) would be expected to be applicable only to those liquids which obey the Maxwell model of viscoelasticity (Section 5.4). In other liquids agreement with experiment is poor, particularly at low temperatures where the temperature dependence of G_∞ is greatly overestimated (Barlow and coworkers, 1966; Madigosky, McDuffie and Litovitz, 1967).

Equation (9.3.6) suggests that

$$J_\infty = 1/G_\infty \propto V_f \qquad (9.3.7)$$

where V_f is the free volume in the liquid. Barlow and coworkers (1966) argued that the temperature dependence of J_∞ is intimately related to free volume. At the temperature T_0 at which there is no expansion free volume, J_∞ should have a value $(J_\infty)_0$ which is characteristic of the equilibrium glassy liquid. As the temperature of the liquid is increased, J_∞ also increases because of the loosening of the liquid structure on account of the increase in free volume. Barlow and coworkers (1966) empirically proposed that

$$J_\infty = (J_\infty)_0 + B'V_f \qquad (9.3.8)$$

where B' is a constant. Since V_f is approximately proportional to $(T - T_0)$, (9.3.8) may be written

$$\frac{1}{G_\infty} = \frac{1}{(G_\infty)_0} + B(T - T_0) \qquad (9.3.9)$$

which is identical with equation (9.2.1) deduced from experiment. Equation (9.3.9) is wholly empirical, however, and attempts to correlate $(G_\infty)_0$ and B with molecular parameters have proved unsuccessful. This is hardly surprising since both these parameters will depend critically on the exact details of the intermolecular potential and on the ability of the molecules to undergo internal rotation and overall molecular rotation on application of a shear stress.

An alternative approach to the calculation of G_∞ of liquids is that of Madigosky, McDuffie and Litovitz (1967) who assume that the compliance of an associated liquid is inversely proportional to the density of hydrogen bonds in the liquid:

$$J_\infty = A + BV/N_H \qquad (9.3.10)$$

A and B are constants, V is the molar volume of the liquid and N_H is the number of hydrogen bonds per mole. At T_0, (9.3.10) becomes

$$(J_\infty)_0 = A + BV_0/(N_H)_0 \qquad (9.3.11)$$

and thus

$$J_\infty = (J_\infty)_0 + B[V/N_H - V_0/(N_H)_0] \qquad (9.3.12)$$

If the number of hydrogen bonds decreases with increasing temperature according to the relation

$$\frac{(N_H)_0 - N_H}{N_H} = \exp(-\Delta G/RT) \qquad (9.3.13)$$

where ΔG is the free energy required to rupture one mole of hydrogen bonds, then

$$J_\infty = (J_\infty)_0 + \frac{BV}{n_H N}\left[\exp\left(-\frac{\Delta H}{RT} + \frac{\Delta S}{R}\right) + \frac{V_f}{V}\right] \qquad (9.3.14)$$

n_H is the number of hydroxyl groups per molecule and N is Avogadro's number. Equation (9.3.14) gives good agreement with experiment, and values of ΔH and ΔS required to describe the pressure and temperature dependence of a number of associated liquids are consistent with values obtained from other sources. If a liquid contains no hydrogen bonds, (9.3.14) reduces to (9.3.8).

Thus present theories of G_∞ of liquids fall into two classes: rigorous statistical mechanical treatments and empirical hypotheses. The former may be evaluated exactly with certain assumptions, but they cannot be tested as no experimental data are available for liquids for which the assumptions are valid. Although the empirical theories can describe the experimental results adequately, they contain a number of adjustable parameters which cannot be calculated from molecular or thermodynamic data.

9.4   The Viscoelastic Relaxation Region of Simple Liquids

The viscoelastic relaxation region of a liquid may be found as a function of frequency by the application of the principle of reduced variables to experiments with shear waves at a small number of frequencies but over a range of temperature. Values of G_∞ in the relaxation region are determined either by extrapolation from actual measured values at lower temperatures or by choosing G_∞ values such that plots of $R_L/(\rho G_\infty)^{\frac{1}{2}}$ or G'/G_∞ versus $\log(\omega\eta_s/G_\infty)$ for each frequency are identical within experimental error: the values of ρ, G_∞ and η_s refer to the temperature at which R_L or G' has been determined. If the assumptions underlying this reduction are valid, the frequency dependence of the viscoelastic parameters of the liquid at constant temperature is obtained.

In liquids such as argon where the viscosity is less than 10^{-3} N s m^{-2}, theoretical studies suggest that the shear relaxation time is about 10^{-13} s

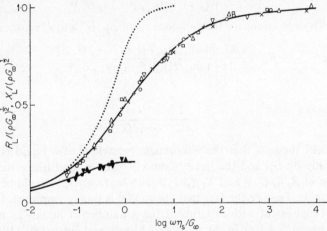

FIGURE 9.4.1. Plots of $R_L/(\rho G_\infty)^{\frac{1}{2}}$ and $X_L/(\rho G_\infty)^{\frac{1}{2}}$ against $\log(\omega\eta_s/G_\infty)$: $\triangledown$, $\blacktriangledown$, squalane; $\square$, $\blacksquare$, tri(β-chloroethyl)phosphate; $\bigcirc$, $\bullet$, tri(m-tolyl)phosphate; $\times$, tris(2-ethyl hexyl)phosphate; $+$, tetra(2-ethyl hexyl)silicate; $\triangle$, $\blacktriangle$, bis(m-(m-phenoxy phenoxy)phenyl)ether; ——, predictions of B.E.L. model, equation (9.4.2); ...., predictions of Maxwell model, equation (5.4.13) (after Barlow, Erginsav and Lamb, 1967).

(Mountain and Zwanzig, 1966; Dexter and Matheson, 1970). Hence the available experimental techniques (Chapter 6) do not generate shear waves of sufficiently high frequency to observe viscoelastic relaxation in such liquids. In more viscous liquids the most extensive results available are those of Barlow, Erginsav and Lamb (1967). Figure 9.4.1 shows their results for a number of liquids obtained using shear waves of 30 MHz frequency and using equation (9.2.1) to extrapolate G_∞ into the relaxation region. The dotted line which corresponds to a single Maxwell relaxation time (equation 5.4.13) is clearly not in agreement with the experimental results. A significant feature of these results is that the normalized components of Z_L are identical within experimental error for each liquid when plotted against a reduced frequency of $\omega\eta_s/G_\infty$. Similar behaviour had earlier been observed in a series of lubricating oils (Barlow and Lamb, 1959; Hutton, 1968).

Barlow, Erginsav and Lamb (1967) conclude that molecular parameters play no part in determining the shape of the viscoelastic relaxation curves of a simple liquid. Instead, they suggest that the shear mechanical impedance of a viscoelastic liquid should be represented by a parallel combination of the impedances of a Newtonian liquid and a Hookean solid:

$$1/Z_L = 1/(1+i)(\pi f \eta_s \rho)^{\frac{1}{2}} + 1/(\rho G_\infty)^{\frac{1}{2}} \qquad (9.4.1)$$

This yields expressions for the components of Z_L, G^* and J^*. For example

$$R_L = \frac{(\rho G_\infty)^{\frac{1}{2}}(\omega\eta_s/2G_\infty)^{\frac{1}{2}}[1+(2\omega\eta_s/G_\infty)^{\frac{1}{2}}]}{[1+(\omega\eta_s/2G_\infty)^{\frac{1}{2}}]^2+(\omega\eta_s/2G_\infty)} \qquad (9.4.2)$$

and

$$J' = \frac{1}{G_\infty} + \frac{1}{(\omega\eta_s G_\infty/2)^{\frac{1}{2}}} \qquad (9.4.3)$$

This model suggests that the viscoelastic behaviour of a liquid is determined only by G_∞ and the steady flow viscosity η_s, and hence the same variation of $R_L/(\rho G_\infty)^{\frac{1}{2}}$ and $X_L/(\rho G_\infty)^{\frac{1}{2}}$ with $\log(\omega\eta_s/G_\infty)$ should be found in all liquids. Despite the fact that the model is purely empirical, there is excellent agreement with experiment for a number of liquids as may be seen in Figure 9.4.1 or in the results of Moore and coworkers (1969) for the viscoelastic relaxation of pentachlorodiphenyl.

The Barlow–Erginsav–Lamb (B.E.L.) model is less successful when applied to step-function experiments: the expression for the creep compliance is (Erginsav, 1969)

$$J(t) = J_\infty + t/\eta_s + 4(t/\pi\eta_s G_\infty)^{\frac{1}{2}} \qquad (9.4.4)$$

In addition to the elastic term J_∞ and the viscous term t/η_s (cf. equation 5.6.7), there is an additional term in $t^{\frac{1}{2}}$ which arises from the viscoelastic relaxation processes. Such unlimited creep strain resulting from relaxation is wholly unreal.

The B.E.L. model of equation (9.4.1) is not consistent with many of the available results for the viscoelastic properties of liquids. Barlow, Erginsav and Lamb (1969) investigated a number of binary mixtures of liquids which in their pure form obeyed equation (9.4.1) exactly. They found that if the components of the mixture were similar in molecular weight, then no deviations from the B.E.L. model occurred. In contrast, when the components of the mixture differed significantly in molecular weight, the width of the viscoelastic relaxation region varied widely with the composition of the mixture. In order to describe their results mathematically, Barlow, Erginsav and Lamb (1969) made an empirical modification of their model by adding a parameter K to the expression for J' in equation (9.4.3):

$$J' = \frac{1}{G_\infty} + \frac{K}{(\omega \eta_s G_\infty / 2)^{\frac{1}{2}}} \qquad (9.4.5)$$

K is a measure of the width of the distribution of relaxation times, the larger K values corresponding to a broader distribution: $K = 1$ represents the unmodified B.E.L. model, while $K = 0$ is the Maxwell model (Section 5.4). Figure 9.4.2 shows the variation of the parameter K with mole fraction of the lower molecular weight component for a number of liquid mixtures. A feature of this curve is that a number of different

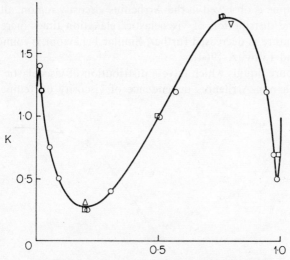

Mole fraction lower molecular weight component

FIGURE 9.4.2. Variation of the parameter K in equation (9.4.5) with mole fraction for a number of liquid mixtures: $\circ$, tri(m-tolyl)phosphate and di(n-butyl)phthalate; $\square$, tetra(2-ethyl hexyl)silicate and squalane; $\triangle$, bis(m-(m-phenoxy phenoxy)phenyl)ether and tri(β-chloroethyl)phosphate; ∇, tri(m-tolyl)phosphate and tri(β-chloroethyl)phosphate (after Barlow, Erginsav and Lamb, 1969).

liquid mixtures show the same K variation. A few results which are consistent with those just quoted are those of Knollman, Miles and Hamamoto (1965) on hexachlorodiphenyl where 1% of the methanol or toluene impurity broadens the distribution significantly. Slie, Donfor and Litovitz (1966), however, have found that the breadth of the viscoelastic relaxation region of glycerol–water mixtures is independent of water concentration in the range 0–68 mole per cent water.

As pointed out previously, the viscoelastic relaxation region of a liquid is usually defined by applying the principle of reduced variables to measurements made over a wide range of temperature and narrow range of frequency. Moreover, the liquid viscosity in the temperature range of interest is normally described by equation (9.1.2) and not by the Arrhenius equation (9.1.1). In liquid boron trioxide and zinc chloride, however, the Arrhenius viscosity region extends to sufficiently high viscosities for viscoelastic relaxation to be observed with shear wave frequencies below 100 MHz. Figure 9.4.3 shows the viscoelastic relaxation region of B_2O_3 as determined by Tauke, Litovitz and Macedo (1968) at a number of constant temperatures with a range of shear wave frequencies: a single Maxwell relaxation time is observed in the Arrhenius viscosity region, although the width of the distribution of viscoelastic relaxation times increases when the temperature is decreased further. Similar behaviour is found in $ZnCl_2$ (Gruber and Litovitz, 1964).

Not all pure liquids which have a distribution of viscoelastic relaxation times and a non-Arrhenius dependence of viscosity on temperature are

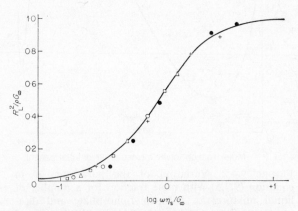

FIGURE 9.4.3. Normalized square of the shear mechanical resistance of boron trioxide as a function of reduced frequency: ●, 1073 K; +, 1123 K; △, 1173 K; □, 1223 K; ○, 1273 K; —, single relaxation equation (5.4.13) (after Tauke, Litovitz and Macedo, 1968).

consistent with the unmodified B.E.L. model of viscoelasticity. Litovitz and coworkers have found a wide range of distributions of relaxation times in associated liquids, and have fitted these to Gaussian or Cole–Davidson distributions of relaxation times. (Section 5.5). Some representative results are given by Litovitz and Davis (1965). Similar variations have been found by Davies (1968) for a number of hydrocarbons and their derivatives. In general, the more nearly the temperature dependence of viscosity approaches the Arrhenius relation, the narrower the distribution of viscoelastic relaxation times (Kono, McDuffie and Litovitz, 1966).

The B.E.L. model (with variable parameter K) or the Gaussian and Cole–Davidson functions give convenient mathematical descriptions of the available experimental results for the viscoelastic relaxation region of simple liquids. Various attempts have been made to predict the width of the viscoelastic relaxation region. The first approach is based on the suggestion of Fröhlich (1949) and Kauzmann (1948) that a distribution of relaxation times is the result of a distribution of activation energies for movement of molecules in a liquid. Such a distribution is attributed to the different environments which surround different molecules in the liquid as a result of random thermal fluctuations in the liquid structure. If translational motion occurs non-cooperatively by individual liquid molecules jumping into vacant sites in the liquid (Eyring, 1936), a distribution of environments should lead to a distribution of jump times and a distribution of shear relaxation times. This theory predicts that the introduction of impurity molecules or an increase in temperature should broaden the distribution because of a decrease in the environmental similarity of liquid molecules. This is not in agreement with experiment, such broadening being observed only sometimes on addition of small amounts of low molecular weight impurities and not on increasing the temperature. Nevertheless Knollman, Miles and Hamamoto (1965) have shown that environmental dissimilarity does play a minor role in liquid mixtures.

An alternative approach is that of McDuffie and Litovitz (1962) who suggest that the distribution of shear relaxation times is a result of the existence of regions of short-range order in liquids which are continually breaking up and reforming. The break-up of such structures is assumed to be cooperative so that the size of a given structure decreases non-exponentially with time. If translational motion of an individual molecule can occur only when it has 'melted' from the cooperative structure, then the distribution of lifetimes of molecules in the cooperative structure should lead to a distribution of relaxation times. As the temperature of the liquid increases, the structures decrease in size until a single viscoelastic relaxation time occurs when the lifetime of the clusters is comparable to the time between diffusional jumps of an individual molecule. This hypothesis is in good qualitative agreement with the results for dielectric and viscoelastic relaxation of simple liquids although it does not permit quantitative predictions since little information is available on the size or nature of the ordered regions.

A third attempt to explain the distribution of viscoelastic relaxation times is that of Davies (1968) who suggested that it is determined by the ease of molecular rotation. In the Arrhenius region of liquid viscosity where molecules can rotate many times during the time between translational jumps, the probability of such a jump is not dependent on molecular orientation and a single Maxwell relaxation time is found. Liquids containing non-spherical molecules show a non-Arrhenius viscosity region on cooling to lower temperatures where molecular rotation occurs more slowly than molecular translational motion. As a result the probability of a translational movement by a given molecule will depend on its orientation, and the distribution of orientations should lead to a distribution of viscoelastic relaxation times. Internal molecular conformational changes will also be important: if the barrier to such a change is less than the activation energy required for translational motion of the unperturbed molecule, then the molecule will change its conformation into one requiring a lower activation energy for flow from its particular environment. Hence the more spherical a molecule or the more readily it can attain sphericity by a conformational change, the narrower will be the distribution of viscoelastic relaxation times.

There have been several attempts to relate the viscoelastic behaviour of liquids to molecular diffusion. Isakovich and Chaban (1966a, b) have regarded a liquid as consisting of two phases, a long-lived ordered region interspersed with a disordered phase. At equilibrium both the ordered and disordered phases have a definite concentration of holes. When a liquid is disturbed by the propagation of a shear wave, the ordered regions are rearranged and the number of holes changed. The position of equilibrium is then restored by diffusion of molecules across the interfaces between the ordered and disordered regions. For shear deformations there is assumed to be an additional mechanism of Maxwellian relaxation (Section 5.4). These two processes—diffusion and relaxation—give rise to a distribution of shear relaxation times which is in good agreement with experiment in a number of liquids. The difficulty of the Isakovich–Chaban model is that it postulates the existence of long-lived, quasi-crystalline structures in the liquid which have not been detected experimentally.

The Glarum (1960) defect diffusion model of relaxation has been applied to viscoelastic relaxation in liquids. Glarum assumed that motion of a molecule at a particular site in a liquid is only possible when a 'defect' or hole arrives at that site. The motion of a defect through the liquid is described by a diffusion equation, and when a defect reaches a site instantaneous relaxation of the molecule at that site occurs. Although the Glarum model is in good agreement with the results of dielectric and nuclear magnetic relaxation (Hunt and Powles, 1966), it is not directly applicable to viscoelastic relaxation since it does not allow viscous flow of the liquid. Phillips (1969) combined the Glarum model with an independent single relaxation time describing the flow of the liquid. He then assumed that either the stresses or the strains resulting from the two mechanisms were additive. The addition of stresses together with some mathematical approximations gave the modified Barlow, Erginsav and Lamb (1969) model of viscoelasticity (equation 9.4.5): the addition of the strains gave the B.E.L. model exactly when $\omega \eta_s / G_\infty > 1$. The parameter K in equation (9.4.5) is given by

$$2K = (\tau_f / \tau_d)^{\frac{1}{2}} \qquad (9.4.6)$$

where τ_f is the relaxation time of the flow process and τ_d is the time constant of the diffusion. Since the dependence on temperature of τ_f and τ_d will differ, the width of the viscoelastic relaxation region is temperature dependent. A similar approach has been adopted by Montrose and Litovitz (1970).

Thus a number of different models are able to account qualitatively for the viscoelastic behaviour of viscous liquids, models involving diffusion being the most successful. A quantitative prediction of the behaviour of a given liquid is not yet possible. In the diffusion models, for example, the relative importance of the spontaneous relaxation process and the diffusion process can only be determined experimentally. It should again be emphasized that most experimental work on viscoelastic relaxation involves the principle of reduced variables. Since this is not applicable over a wide temperature range as can be seen in the results for B_2O_3 given above, it is probable that small errors are introduced by its application to the modest range of temperature used in viscoelastic studies.

CHAPTER 10

STRUCTURAL RELAXATION IN LIQUIDS

10.1 Introduction

The propagation of a pure shear wave through a liquid, discussed in Chapters 5 and 9, gives information on the rigidity moduli of liquids and on viscoelastic relaxation. The propagation of an alternating compression through a liquid may be treated similarly. From the general stress–strain relation or constitutive equation of the liquid, one can calculate the dependence upon frequency of the compressional velocity and absorption and the compressional elastic constants of the liquid. In the case of a shear wave, if the liquid molecules are unable to flow sufficiently quickly to respond to an applied shear stress of high frequency, the elastic modulus G' increases with frequency while the loss modulus G'' goes through a maximum when $\omega\tau_s \approx 1$ (Section 5.4). With an alternating compression of low frequency, the volume fluctuations are able to remain in phase with the applied pressure variations by the flow of liquid molecules between positions of high and low density. As the compressional frequency increases, however, the liquid structure is not able to adjust itself to the pressure variations sufficiently quickly: structural relaxation occurs, the bulk modulus K' increases while the loss modulus K'' goes through a maximum where $\omega\tau_v \approx 1$. τ_v is the volume or structural relaxation time, that is the time required for a liquid to adjust itself to its new equilibrium volume following a rapid change in the applied pressure.

It is not possible experimentally to study the propagation of pure compressional waves in a liquid. An ultrasonic longitudinal wave contains both shear and compressional components, however, and the propagation of such a wave may be discussed in terms of the shear and compressional moduli and relaxation times. In this Chapter we shall investigate the factors which determine the compressional elastic moduli and structural relaxation times of liquids where two-state equilibria are absent. The latter are discussed in Chapters 11 and 12.

10.2 Response of a Medium to a Compression

In Section 5.3, the response of a viscoelastic medium to a shear stress was characterized by a complex rigidity modulus G^* or a complex viscosity η^*. The stress–strain relation was

$$\frac{\partial \gamma}{\partial t} = \frac{1}{G_\infty}\frac{\partial \sigma}{\partial t} + \frac{\sigma}{\tau_s G_\infty} \tag{5.4.1}$$

or

$$\sigma + \tau_s \frac{\partial \sigma}{\partial t} = G_\infty \tau_s \frac{\partial \gamma}{\partial t} \tag{10.2.1}$$

σ is the shear stress, γ the shear strain, G_∞ the infinite frequency rigidity modulus of the medium, and τ_s the shear relaxation time. If the liquid has a zero frequency rigidity modulus G_0 as in a rubber, (10.2.1) is generalized to

$$\sigma + \tau_s \frac{\partial \sigma}{dt} = G_0 \gamma + G_\infty \tau_s \frac{\partial \gamma}{\partial t} \tag{10.2.2}$$

The response of a medium to a compression may be characterized by a complex bulk modulus K^* or a complex bulk viscosity η_B^*. A stress–strain relation similar to (10.2.2) may be written for compression:

$$P + \tau_B \frac{\partial P}{\partial t} = K_0 \gamma + K_\infty \tau_B \frac{\partial \gamma}{\partial t} \tag{10.2.3}$$

K_0 is the zero frequency bulk modulus which is finite in all liquids, K_∞ is the bulk modulus at infinite frequency, τ_B is the bulk relaxation time, and the strain γ is the fractional volume change.

If we apply a sinusoidally varying pressure

$$P = P_0 \exp(i\omega t) \tag{10.2.4}$$

to a liquid to which equation (10.2.3) applies, we obtain (cf. equation 5.4.9)

$$K^* = K' + iK'' = \frac{P}{\gamma} = \frac{K_0 + i\omega\tau_B K_\infty}{1 + i\omega\tau_B} \tag{10.2.5}$$

The components of K^* are (cf. equation 5.4.10)

$$K' = K_0 + \frac{K_2 \omega^2 \tau_B^2}{1 + \omega^2 \tau_B^2}; \quad K'' = \frac{K_2 \omega \tau_B}{1 + \omega^2 \tau_B^2} \tag{10.2.6}$$

Here $K_2 = K_\infty - K_0$ is the relaxational part of the bulk modulus. The frequency dependence of $(K' - K_0)$ and K'' is shown in Figure 10.2.1: it is the same as that of G' and G'' in Figure 5.4.1.

The complex bulk viscosity η_B^* may be defined by (cf. equation 5.3.9)

$$\eta_B^* = \eta_B' - i\eta_B'' = K^*/i\omega \tag{10.2.7}$$

Hence

$$\eta_B' = \frac{K''}{\omega} = \frac{K_2 \tau_B}{1 + \omega^2 \tau_B^2} \tag{10.2.8}$$

At low frequencies ($\omega\tau_B \ll 1$), η_B' equals the bulk viscosity η_B and

$$\eta_B = \tau_B K_2 \tag{10.2.9}$$

an expression analogous to the Maxwell relation (5.4.6).

The expressions for K^* and η_B^* are similar to the expressions for G^* and η_s^* given in Chapter 5. Each of the shear functions, including the shear mechanical impedance Z_L, the complex compliance J^*, the stress relaxation function $\psi(t)$ and the creep compliance $J(t)$ has its equivalent for compression: moreover, the compressional functions may be inter-converted in the same way as for the shear functions (Gross, 1968;

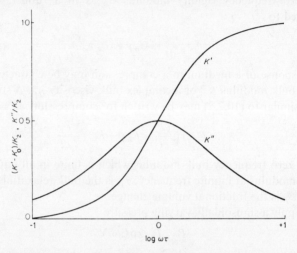

FIGURE 10.2.1. Frequency dependence of the com-ponents $(K' - K_0)$ and K'' of the relaxational part K_2 of the bulk modulus of a fluid having a single struc-tural relaxation time.

Schwarzl and Struik, 1968). Usually discussions of compressional relaxa-tion are confined to the functions K^* and η_B^*, although the complex compressibility β^* ($=1/K^*$) is sometimes used (Section 12.3).

Equations (10.2.2), (10.2.6) and (10.2.8) are applicable when a single relaxation time is able to describe the experimental results. As in the case of shear relaxation (Chapter 5), this description is often found experi-mentally to be inadequate, and it is convenient mathematically to represent the results by a distribution of relaxation times. Equation (10.2.6) can then be generalized to give

$$\left.\begin{aligned} K' &= K_0 + K_2 \int_0^\infty F(\tau_B) \frac{\omega^2 \tau_B^2}{1 + \omega^2 \tau_B^2}\, d\tau_B \\[2mm] K'' &= K_2 \int_0^\infty F(\tau_B) \frac{\omega \tau_B}{1 + \omega^2 \tau_B^2}\, d\tau_B \end{aligned}\right\} \qquad (10.2.10)$$

$F(\tau_B)$ represents the distribution of structural relaxation times.

$K_2 F(\tau_B) \, d\tau_B$ is the contribution to the relaxational bulk modulus of the medium from those processes which have relaxation times in the range τ_B to $(\tau_B + d\tau_B)$. $F(\tau_B)$ is normalized such that

$$\int_0^\infty F(\tau_B) \, d\tau_B = 1 \qquad (10.2.11)$$

The bulk viscosity of the medium is then

$$\eta_B = K_2 \int_0^\infty \tau_B F(\tau_B) \, d\tau_B = K_2 \bar{\tau}_B \qquad (10.2.12)$$

where $\bar{\tau}_B$ is an average structural relaxation time (cf. equation 10.2.9).

Expressions may also be derived for the components of β^* in terms of the distribution of structural retardation times (cf. equation 5.5.7).

It should be noted that K^*, τ_B, η_B^*, etc. may be either isothermal or adiabatic depending upon the particular experiment. When the compression results from the passage of an ultrasonic longitudinal wave through the liquid, these parameters assume their adiabatic values.

10.3 Propagation of an Ultrasonic Longitudinal Wave through a Relaxing Medium

Since it is not possible experimentally to propagate a pure compressional wave in a liquid, information on structural relaxation is obtained by studying the propagation of an ultrasonic longitudinal wave. The equation describing the propagation of a longitudinal wave is (cf. equations 2.1.4 and 5.3.3)

$$\rho \frac{\partial^2 \xi}{\partial t^2} = M^* \frac{\partial^2 \xi}{\partial x^2} \qquad (10.3.1)$$

ρ is the density of the medium, ξ the particle displacement, and $M^* = M' + iM''$ is the complex longitudinal elastic modulus. The complex velocity of the longitudinal wave is

$$U^* = (M^*/\rho)^{\frac{1}{2}} \qquad (10.3.2)$$

M^* contains contributions from the compressional and shearing motions associated with a longitudinal wave and can be written (Herzfeld and Litovitz, 1959)

$$M^* = K^* + \tfrac{4}{3} G^* \qquad (10.3.3)$$

The solution of (10.3.1) is (cf. equations 2.1.6 and 5.3.5)

$$\xi = \xi_0 \exp \left[i\omega \left\{ t - x \left(\frac{1}{U} + \frac{\alpha}{i\omega} \right) \right\} \right] \qquad (10.3.4)$$

$$= \xi_0 \exp \left[i\omega (t - x/U^*) \right] \qquad (10.3.5)$$

U is the phase velocity, α the absorption coefficient and

$$\frac{1}{U^*} = \frac{1}{U} - \frac{i\alpha}{\omega} \tag{10.3.6}$$

The components of M^* may be determined in terms of α and U by combining equations (10.3.2) and (10.3.6) to give

$$M' = \rho U^2 \left[\frac{1 - (\alpha U/\omega)^2}{\{1 + (\alpha U/\omega)^2\}^2} \right] \tag{10.3.7}$$

and

$$M'' = \frac{2\rho U^2 (\alpha U/\omega)}{\{1 + (\alpha U/\omega)^2\}^2} \tag{10.3.8}$$

Hence M' and M'' can be calculated from the observed U and α values. If G' and G'' are known from shear wave studies, K' and K'' may be obtained from equation (10.3.3).

If the absorption coefficient of the sound wave is small so that $(\alpha U/\omega)^2 \ll 1$, (10.3.7) and (10.3.8) reduce to

$$M' = \rho U^2 \tag{10.3.9}$$

and

$$M'' = 2\rho U^3 \alpha/\omega \tag{10.3.10}$$

Hence

$$U^2 = \frac{M'}{\rho} = \frac{1}{\rho}[K' + \tfrac{4}{3}G']$$

$$= \frac{1}{\rho}\left[K_0 + \frac{K_2 \omega^2 \tau_B^2}{1 + \omega^2 \tau_B^2} + \frac{\tfrac{4}{3}G_\infty \omega^2 \tau_s^2}{1 + \omega^2 \tau_s^2} \right] \tag{10.3.11}$$

by (10.2.6) and (5.4.10). Similarly (10.3.10) gives an expression for the absorption per wavelength

$$\mu = \alpha\lambda = \frac{\pi M''}{\rho U^2} = \frac{\pi}{\rho U^2}(K'' + \tfrac{4}{3}G'')$$

$$= \pi \left[\frac{\dfrac{K_2 \omega \tau_B}{1 + \omega^2 \tau_B^2} + \dfrac{\tfrac{4}{3}G_\infty \omega \tau_s}{1 + \omega^2 \tau_s^2}}{K_0 + \dfrac{K_2 \omega^2 \tau_B^2}{1 + \omega^2 \tau_B^2} + \dfrac{\tfrac{4}{3}G_\infty \omega^2 \tau_s^2}{1 + \omega^2 \tau_s^2}} \right] \tag{10.3.12}$$

It must be emphasized that (10.3.9) to (10.3.12) are applicable only when $(\alpha U/\omega)^2 \ll 1$. In many liquids of high viscosity, the exact expressions (10.3.7) and (10.3.8) must be used.

At low frequencies ($\omega\tau \ll 1$), (10.3.11) reduces to

$$U_0^2 = K_0/\rho \tag{10.3.13}$$

which is identical to (2.1.11) since $K = 1/\beta$. At high frequencies ($\omega\tau \ll 1$)

$$U_\infty^2 = \frac{1}{\rho}[K_0 + K_2 + \tfrac{4}{3}G_\infty]$$ (10.3.14)

At low frequencies, (10.3.12) reduces to

$$\alpha\lambda = \frac{\pi\omega}{K_0}(K_2\tau_B + \tfrac{4}{3}G_\infty\tau_s)$$

$$= \frac{\pi\omega}{K_0}(\eta_B + \tfrac{4}{3}\eta_s)$$ (10.3.15)

using (10.2.9) and (5.4.6). Equation (10.3.15) is equivalent to (2.2.8). If the velocity dispersion is small and $K_0 \gg K_2 + G_\infty$, equation (10.3.12) becomes

$$\alpha\lambda \approx \frac{\pi\omega}{K_0}(\eta'_B + \tfrac{4}{3}\eta'_s)$$ (10.3.16)

10.4 The Bulk Viscosity of Liquids

The bulk viscosity η_B of a liquid may be calculated from the observed ultrasonic absorption by the relation

$$\alpha_{obs} - \alpha_{cl} = \frac{2\pi^2 f^2}{\rho U^3}\eta_B$$ (2.2.8)

α_{cl} is the ultrasonic absorption caused by both the shear viscosity and the thermal conductivity of the liquid, although with the exception of liquid metals the contribution from thermal conductivity is only a few per cent of α_{cl}.

All liquids have a bulk viscosity, and Table 10.4.1 gives some results for the shear and bulk viscosities of some monatomic and associated liquids. The ratio of η_B to η_s is approximately independent of temperature and usually lies between 1 and 3. The decrease in η_s and η_B with increasing temperature generally leads to a decrease in α/f^2, although at high temperatures, where ρU^3 in equation (2.2.8) decreases more rapidly than η_B and η_s, α/f^2 increases slowly with temperature (Piercy and Rao, 1967b).

A similar decrease in η_B and α/f^2 with increasing temperature is shown by non-associated liquids at low temperatures as may be seen from the results for toluene plotted in Figure 10.4.1 (Piercy and Rao, 1967b). At high temperatures, however, α/f^2 increases rapidly with temperature so that η_B/η_s in toluene increases from $1\cdot7 \pm 0\cdot2$ at 195 K to an apparent value of several thousand near the liquid boiling point. This high temperature behaviour is the result of the increasing vibrational heat capacity of the liquid. Vibrational relaxation in liquids (Chapter 11) is normally so rapid that the entire vibrational heat capacity contributes to the propagation of sound waves of frequencies below 100 MHz in liquids. The increase in vibrational heat capacity with increasing temperature leads to an increase

TABLE 10.4.1. Shear and bulk viscosities of liquids

Liquid	T (K)	P (atm)	η_s (N s m^{-2})	η_B (N s m^{-2})	η_B/η_s	Reference*
Argon	87	19	2.56×10^{-4}	1.96×10^{-4}	0.76	1
	106	32	1.50×10^{-4}	1.95×10^{-4}	1.30	1
	131	26	7.3×10^{-5}	1.84×10^{-4}	2.52	1
Sodium	377	1	6.8×10^{-4}	1.8×10^{-3}	2.6	2
Mercury	298	1	1.6×10^{-3}	1.9×10^{-3}	1.2	3
	477	1	1.0×10^{-3}	1.3×10^{-3}	1.3	3
Bismuth	578	1	1.7×10^{-3}	7.0×10^{-3}	4.2	3
	715	1	1.3×10^{-3}	5.6×10^{-3}	4.3	3
Nitrogen	70.2	0.4	2.2×10^{-4}	1.4×10^{-4}	0.63	4
	75.3	0.8	1.7×10^{-4}	0.8×10^{-5}	0.46	4
Oxygen	66.3	0.03	4.2×10^{-4}	3.9×10^{-4}	0.94	4
	86.1	0.6	2.1×10^{-4}	2.0×10^{-4}	0.96	4
Water	288	1	1.1×10^{-3}	3.1×10^{-3}	2.8	5
Methanol	303	1	5.2×10^{-4}	8.4×10^{-4}	1.6	6
Ethanol	303	1	1.0×10^{-3}	1.4×10^{-3}	1.4	6
n-Propanol	303	1	1.9×10^{-3}	2.2×10^{-3}	1.2	6
n-Butanol	303	1	2.3×10^{-3}	2.4×10^{-3}	1.1	6
Glycerol	327	1	1.1×10^{-1}	1.1×10^{-1}	1.0	7
Sodium nitrate	583	1	2.9×10^{-3}	1.1×10^{-2}	3.9	8

* References
1. Naugle, Lunsford and Singer (1966). 5. Pinkerton (1947).
2. Jarzynski and Litovitz (1964).   6. Carnevale and Litovitz (1955).
3. Jarzynski (1963).   7. Piccirelli and Litovitz (1957).
4. Victor and Beyer (1970).   8. Higgs and Litovitz (1960).

in the value of α/f^2 at frequencies well below the vibrational relaxation region as may be seen in Figure 11.2.2. From an analysis of the data available on the vibrational heat capacities of non-associated liquids, Piercy and Rao (1967b) showed that the temperature dependence of the vibrational contribution to the sound absorption could be adequately represented by the empirical relation

$$(\alpha/f^2)_{\text{vib}} = A + BT \qquad (10.4.1)$$

where A and B are constants. When the vibrational contribution for toluene (VIB in Figure 10.4.1) is subtracted from the observed α/f^2, then the contribution from viscosity to the absorption is obtained (VISC in Figure 10.4.1). The temperature dependence of $(\alpha/f^2)_{\text{visc}}$ is now the same for associated and unassociated liquids, while the ratio of the bulk viscosity with the vibrational contribution removed to the shear viscosity is now independent of temperature over the entire temperature range in non-associated liquids.

Thus the temperature dependence of the observed (α/f^2) and bulk viscosity of liquids depends on the relative magnitudes of the viscous

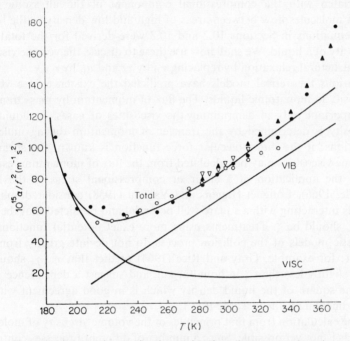

FIGURE 10.4.1. Ultrasonic absorption in toluene as a function of temperature: VISC, contribution of shear and volume viscosities; VIB, contribution of internal molecular vibrations: Total, sum of VISC and VIB: ●, Piercy and Rao (1967b), 15 MHz; ▲, Piercy and Rao (1967b), 45 MHz; □, Heasell and Lamb (1956b), 100 MHz; ○, Nozdrev (1965); ▽, Cerceo, Meister and Litovitz, (1962) (after Piercy and Rao, 1967b).

and vibrational contributions to the ultrasonic absorption. In associated liquids at room temperature and below and in non-associated liquids at low temperatures, the absorption caused by the high shear and volume viscosities is much greater than the vibrational contribution. At higher temperatures in non-associated liquids, the converse is true. Rotational isomerization and chemical equilibria can also lead to high ultrasonic absorptions and anomalous bulk viscosities. When allowance has been made for such effects, however, the ratio of the 'true' volume viscosity to the shear viscosity appears to lie between 1 and 3 for most liquids and to be independent of temperature within experimental error (Litovitz and Davis, 1965).

In the remainder of this Chapter we shall consider only this 'true' volume viscosity or structural viscosity. It represents the energy loss associated with the compressional component of the ultrasonic wave when molecules flow between areas of high and low density in the liquid. The equations in Sections 10.2 and 10.3 were derived for the total bulk viscosity of a liquid. We shall now use these to discuss the volume viscosity and structural relaxation by replacing τ_B by τ_v and η_B by η_v.

Various theoretical models have predicted the existence of a volume viscosity in monatomic liquids. The flux of momentum by mass transport is important only in determining the viscosities of gases. In liquids the viscosity is determined by the transfer of momentum during molecular collisions: if the intermolecular force function is known, the shear and volume viscosities may be calculated from the flux of momentum resulting from the application of a shear or compressional stress to the liquid (Pryde, 1966). Longuet-Higgins and Valleau (1958) considered spherical atoms interacting with a square-well potential and predicted that the ratio n_v/η_s should be $\frac{5}{3}$. Treatments using more exact potential functions and realistic models of the collision process do not deviate greatly from this result: for example, Gray and Rice (1964) predict that η_v/η_s should be only slightly less than $\frac{5}{3}$ in liquid argon, and suggest a dependence of η_v on the square of the liquid density which is in good agreement with the experimental results in Table 10.4.1.

The calculation from first principles of the volume viscosity of molecular liquids is not yet possible. Since a number of different processes contribute to the observed bulk viscosity of liquids, even an accurate experimental value of the volume viscosity cannot always be obtained. This causes difficulties in treating the structural relaxation region of liquids (Section 10.6), and it is assumed that the ratio η_v/η_s is independent of temperature. While this is a reasonable assumption over a wide range of temperature at low viscosities (Piercy and Rao, 1967b), it cannot be verified in high viscosity liquids where structural relaxation occurs.

10.5 Sound Velocity in Liquids when Relaxation Effects are Absent

Figure 10.5.1 shows the results of Litovitz, Lyon and Peselnick (1954) for the velocity of ultrasonic waves of frequencies between 1 and 50 MHz in pentachlorodiphenyl. A dispersion of the velocity is clearly seen around 300 K, while the sound velocity is independent of frequency at temperatures above and below the relaxation region. In this Section we shall consider the temperature dependence of the ultrasonic velocity U at temperatures above those of the relaxation region (U_0) and below (U_∞).

Many measurements have been made of the ultrasonic velocity U_0 in liquids and of its temperature dependence, and results are tabulated by Schaaffs (1963), Nozdrev (1965) and Landolt-Börnstein (1967). In a wide range of liquids, associated and non-associated, and including molten salts, the ultrasonic velocity decreases linearly with increasing temperature

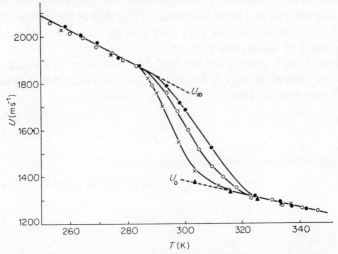

FIGURE 10.5.1. Ultrasonic velocity in pentachlorodiphenyl as a function of temperature: ▲, 1·0 MHz; ×, 7·46 MHz; ○, 21·8 MHz; ●, 50·8 MHz (after Litovitz, Lyon and Peselnick, 1954).

over a considerable range of temperature. Deviations from this linear relation occur near the boiling point and melting point of liquids, so that extrapolation of the measured ultrasonic velocities from higher temperatures to the melting point can lead to errors of the order of -2 to -6% (Nozdrev, 1965). A prominent exception to this behaviour is water where the ultrasonic velocity increases from 273 K to a maximum value at 347 K, and thereafter decreases with increasing temperature (Barlow and Yazgan, 1966).

There have been many attempts to calculate the ultrasonic velocity in a liquid from molecular parameters. For example, Rao (1940, 1941) observed that the approximately linear dependences of sound velocity U_0 and molar volume V of liquids were related by the expressions

$$\frac{1}{U_0}\frac{dU_0}{dT}\bigg/\frac{1}{V}\frac{dV}{dT} = -3 \qquad (10.5.1)$$

or

$$U_0^{\frac{1}{3}} V = U_M \tag{10.5.2}$$

U_M is called the molar velocity or Rao's constant, which is independent of temperature and contains additive contributions from the various atomic groupings within the liquid molecules. A more detailed treatment of ultrasonic velocity in terms of group contributions has been developed by Schaaffs (1963). Nozdrev (1965) has found that the ratio of the observed ultrasonic velocity at a given temperature T to that at the critical temperature T_c when plotted against T/T_c gives a single curve to within $\pm 4\%$ for a wide range of liquids and mixtures.

Although such approaches are useful in providing an estimate of the ultrasonic velocity in a given liquid, they are not likely to lead to accurate quantitative results. This may be seen by rearranging

$$U_0^2 = \frac{1}{\rho \beta_S} \tag{2.1.11}$$

using the well-known thermodynamic relations

$$\beta_T - \beta_S = T \theta^2 / \rho C_p \tag{10.5.3}$$

and

$$\beta_T / \beta_S = C_p / C_v \tag{10.5.4}$$

to give

$$U_0^2 = \frac{C_p (C_p - C_v)}{T \theta^2 C_v} \tag{10.5.5}$$

θ is the thermal expansion coefficient of the liquid. Thus U_0 may be calculated from the measurable quantities C_p, T and θ if C_v is known. Although the factors determining C_v of a liquid are well known (Staveley, Hart and Tupman, 1953), quantitative calculation of C_v is not possible for any but the simplest liquids. Hence U_0 cannot be obtained from molecular parameters. Indeed, measured values of U_0 are often used to calculate C_v of a liquid (Dexter and Matheson, 1968; Rowlinson, 1969).

At low temperatures below the structural relaxation region of liquids, the ultrasonic velocity U_∞ is again independent of frequency as may be seen from the results for pentachlorodiphenyl below 285 K in Figure 10.5.1. Few direct measurements have been made of U_∞ in this region because the pulse technique (Section 3.7) which relies on the movement of delay lines through the liquid cannot readily be used at the high liquid viscosities ($> 10^5$ N s m^{-2}) which exist here. Although measurements of the longitudinal impedance ρU_∞ of the liquid can be made by a normal incidence method (Section 3.7), the velocity can only be obtained to $\pm 1\%$. Ideally the phase comparison technique (Section 3.7) should be used in this region.

At temperatures below the ultrasonic relaxation region the absorption is sufficiently small for equation (10.3.9) to be applicable. The longitudinal modulus M_∞ is then given by

$$\rho U_\infty^2 \approx M_\infty = K_\infty + \tfrac{4}{3}G_\infty = K_0 + K_2 + \tfrac{4}{3}G_\infty \qquad (10.5.6)$$

Thus U_∞ depends on three factors, K_0 the low frequency bulk modulus of the liquid, K_2 the infinite frequency value of the relaxing bulk modulus K' and G_∞ the shear modulus. M_∞ can be determined to about $\pm 2\%$ and G_∞ to $\pm 3\%$ from shear wave measurements, while K_0 is obtained by a linear extrapolation of U_0 to low temperatures and is likely to be subject to a similar error. Since K_2 is only some 20% of M_∞, values obtained for K_2 must be viewed with caution.

Some typical results for M_∞, K_∞ and K_2 are given in Table 10.5.1. The ratio K_2/G_∞ varies from 0·5 to 2·1: in view of the large uncertainty in the K_2 values, one can only conclude that K_2 is approximately equal to G_∞ in molecular liquids. The ratio K_∞/G_∞ is typically between 3 and 4. This may be compared with Cauchy's (1882) identity which relates the shear and bulk moduli of an isotropic material in which the particles interact by two-body central forces:

$$K_\infty = \tfrac{5}{3}G_\infty \qquad (10.5.7)$$

Zwanzig and Mountain (1965) have also derived an expression for K_∞ of such a liquid:

$$K_\infty = \tfrac{2}{3}\rho kT + P + \frac{2\pi}{9}\rho^2 \int_0^\infty dr\, g(r) r^3 \frac{d}{dr}\left[r\frac{d\phi}{dr}\right] \qquad (10.5.8)$$

where P is the pressure, $g(r)$ is the radial distribution function, r is the distance between the centres of two particles, and ϕ is the intermolecular potential energy function. Comparison of (10.5.8) with the corresponding expression for G_∞ (9.3.1) leads to a generalization of the Cauchy identity

$$K_\infty = \tfrac{5}{3}G_\infty + 2(P - \rho kT) \qquad (10.5.9)$$

The extension of (10.5.7) to molecular liquids is not possible. Although the same factors must determine K_∞ and G_∞ in such liquids (cf. Section 9.3), not even a qualitative prediction of the relation between G_∞ and K_∞ can be made at present for molecular liquids.

Although K_2 decreases with increasing temperature, its temperature dependence cannot be established with any accuracy. The temperature dependence of K_∞ is better known, and within experimental error is the same as the temperature dependence of G_∞ in the same liquid (Capps and

TABLE 10.5.1. Elastic moduli of liquids

Liquid	T (K)	M_∞ (10^9 N m^{-2})	G_∞ (10^9 N m^{-2})	K_∞ (10^9 N m^{-2})	K_0 (10^9 N m^{-2})	K_2 (10^9 N m^{-2})	K_2/G_∞	K_∞/G_∞	Reference*
Argon[a]	83·8	3·76	1·23	2·12	1·05	1·07	0·87	1·72	1
sec-Butyl benzene	140	7·2	1·34	5·4	3·9	1·5	1·1	4·0	2, 3
3-Phenyl propyl chloride	160	7·7	1·37	5·9	4·6	1·3	0·9	4·5	2, 3
Bis(m-m-phenoxy phenoxy)phenyl)ether	263	6·5	1·02	5·1	3·7	1·4	1·4	5·0	4, 5
Pentachlorodiphenyl	303	4·5	0·84	3·4	2·8	0·5	0·6	4·0	6, 7
n-Propanol	143	4·3	0·62	3·5	3·2	0·3	0·5	5·6	8, 9
Glycerol	259	12·8	3·00	8·8	5·1	3·7	1·2	2·9	10
Hexanetriol	265	9·5	2·29	6·4	3·7	2·7	1·2	2·8	11
Boron trioxide	973	7·4	1·73	5·1	1·5	3·6	2·1	2·9	12, 13

[a] Calculated values.

* References

1. Dexter and Matheson (1970).
2. Dexter (1970).
3. Davies (1968).
4. Barlow, Erginsav and Lamb (1967).
5. Barlow, Lamb and Tasköprülü (1969).
6. Litovitz, Lyon and Peselnick (1954).
7. Moore and coworkers (1969).
8. Kono, Litovitz and McDuffie (1966).
9. Lyon and Litovitz (1956).
10. Picirelli and Litovitz (1957).
11. Meister and coworkers (1960).
12. Macedo and Litovitz (1965).
13. Tauke, Litovitz and Macedo (1968).

coworkers, 1966). Although a linear (Meister and coworkers, 1960) and an exponential (Clark and Litovitz, 1960) decrease of K_∞ with increasing temperature have been proposed, the uncertainties in the data make a linear variation of $1/K_\infty$ with temperature equally probable: this would be consistent with the behaviour of G_∞ discussed in Section 9.3.

In considering structural relaxation in liquids (Section 10.6), it is necessary to assume some form of temperature dependence of K_2 or K_∞ in order to apply the method of reduced variables to the experimental results. It is commonly assumed that the fractional temperature dependence of either K_2 or K_∞ is the same as that of G_∞ (Barlow, Lamb and Tasköprülü, 1969): a distinction between these two possibilities is generally not possible. The theoretical studies of Dexter and Matheson (1970) on argon found that

$$\frac{1}{K_\infty}\frac{dK_\infty}{dT} = \frac{1}{G_\infty}\frac{dG_\infty}{dT} = -0.026 \qquad (10.5.10)$$

$$\frac{1}{K_2}\frac{dK_2}{dT} = -0.032 \qquad (10.5.11)$$

$$\frac{1}{K_0}\frac{dK_0}{dT} = -0.021 \qquad (10.5.12)$$

Thus in argon the fractional temperature dependences of K_∞ and G_∞ are the same. Further experimental work of improved accuracy will be required to verify if this is also true in molecular liquids.

10.6   Structural Relaxation in Liquids

Measurements of both ultrasonic velocity and absorption are required to define the ultrasonic relaxation region completely. The components of M^* can then be calculated from equations (10.3.7) and (10.3.8). Typical results for velocity dispersion were shown in Figure 10.5.1. Figure 10.6.1 gives the results of Dexter (1970) for the ultrasonic absorption in *sec*-butyl benzene at frequencies between 5 and 75 MHz: not all of the high frequency results could be obtained because of the high α values. The relaxation region around 210 K is the result of rotational isomeric relaxation (Chapter 12). When this is allowed for, the viscous contribution to α/f^2 increases with decreasing temperature because of the increase in the shear and volume viscosities of the liquid (Section 10.4). At about 170 K, structural and shear relaxation begins at these frequencies, and α/f^2 decreases with decreasing temperature despite the increase in viscosity.

The components of $\frac{4}{3}G^*$, determined from shear-wave measurements, may be subtracted from M^* to give the compressional moduli $(K_0 + K')$ and K'' as a function of temperature for each frequency of measurement.

As in the case of viscoelastic relaxation discussed in Chapter 9, the entire structural relaxation region cannot be defined at a given temperature using ultrasonic waves of different frequencies because of the narrow frequency range that is usually available experimentally. Thus results at several frequencies over a range of temperature are combined by the

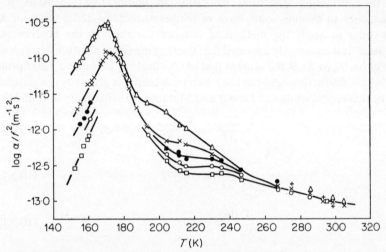

FIGURE 10.6.1. Ultrasonic absorption in *sec*-butyl benzene as a function of temperature: $\triangle$, 5·25 MHz; $\times$, 15·55 MHz; $\bullet$, 25·25 MHz; $\circ$, 35·3 MHz; $\square$, 55·1 MHz; +, 76·2 MHz (after Dexter, 1970).

application of the principle of reduced variables in a manner analogous to that discussed for viscoelastic relaxation in Section (5.7). It is then convenient to plot $(K' - K_0)/K_2$ and K''/K_2 versus $\log \omega \tau_v$ or $\log (\omega \eta_v/K_2)$. This process involves the same assumptions as in the shear wave case, namely that the temperature dependence of all structural relaxation times in a given distribution is the same. η_v is assumed to be a constant multiple of η_s (Section 10.4). Both K_∞ and K_0 are required to calculate K_2, whereas with shear waves only G_∞ is required since G_0 in simple liquids is zero. K_∞ is usually obtained at low temperatures below the structural relaxation region by subtracting $\frac{4}{3}G_\infty$ from the observed M_∞: it is then extrapolated into the relaxation region by assuming a relation between its temperature dependence and that of G_∞ (Section 10.5). K_0 is best determined by a linear extrapolation of the ultrasonic velocity U_0 from higher temperatures (Section 10.5). Some self-consistency can be introduced into this procedure by adjusting the resulting K_2 values to give satisfactory reduction of the data by reduced variables. Nevertheless the values of K' and K'' are subject to the triple uncertainties of the reduced variables assumptions, the extrapolations involved and the errors in subtracting the imperfectly known G^*.

Figure 10.6.2 shows the results of Piccirelli and Litovitz (1957) for $(K' - K_0)/K_2$ of glycerol. As in the case of viscoelastic relaxation in glycerol, the results cannot be described by a single relaxation time and the solid

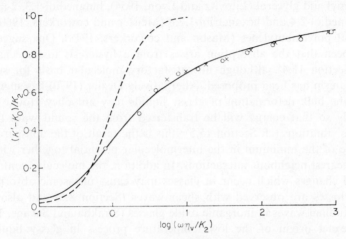

FIGURE 10.6.2. Frequency dependence of the real part of the relaxational bulk modulus of glycerol: o, 22·0 MHz; ×, 51·8 MHz; —, Cole–Davidson distribution of relaxation times with $\beta = 0·32$; ---, single relaxation (after Piccirelli and Litovitz, 1957).

curve in the Figure represents a Cole–Davidson distribution of relaxation times (Section 5.5) with $\beta = 0·32$. The structural relaxation region of several other associated liquids is described by this distribution with different values of β (Meister and coworkers, 1960). In general the distributions of shear and structural relaxation times are not the same (Litovitz and Davis, 1965). In bis(m-(m-phenoxy phenoxy) phenyl) ether, however, Barlow, Lamb and Tasköprülü (1969) find that both shear and structural relaxation can be described by the Barlow, Erginsav and Lamb (1967) model (Section 9.4).

As in viscoelastic relaxation, the width of the distribution of structural relaxation times decreases as the temperature increases. This may be seen by comparing the structural relaxation in glycerol determined at hypersonic frequencies by Brillouin scattering by Pinnow and coworkers (1968) with the ultrasonic results of Piccirelli and Litovitz (1957). The distribution also narrows with increasing temperature in boron trioxide (Macedo and Litovitz, 1965; Tauke, Litovitz and Macedo, 1968) and zinc chloride

(Gruber and Litovitz, 1964) until a single structural relaxation time is found in the Arrhenius viscosity region.

In several liquids at temperatures below the structural relaxation region, an additional process is found which leads to ultrasonic absorption and sometimes to velocity dispersion. This has been observed in pentachloro-diphenyl and glycerol (Litovitz and Lyon, 1954), butanediol-1,3, 2-methyl pentanediol-2,4 and hexanetriol-1,2,6 (Meister and coworkers, 1960) and several polyisobutylenes (Mason and coworkers, 1948). One suggestion has been that the absorption arises from a hysteresis loss mechanism (cf. Section 15.4), although no satisfactory molecular basis for such a mechanism has been proposed. Alternatively Dexter (1970) has suggested that the bulk deformation of glassy liquids may not obey Hooke's law exactly so that energy will be transferred from the sound wave to the lattice vibrations (cf. Section 15.5): this is the result of the non-parabolic nature of the minimum in the intermolecular potential together with the non-nearest neighbour interactions. In addition, the molecular conforma-tional changes which occur in glasses may cause ultrasonic absorption: such losses are observed with shear waves (Section 8.5) and also with longitudinal waves in inorganic oxide glasses (Strakna and Savage, 1964). The exact origin of the low temperature process in glassy liquids is uncertain. Where it overlaps the structural relaxation region, it complicates considerably the interpretation of the latter. As already described, there are considerable experimental errors in obtaining the structural relaxation region which is consequently less well understood than the viscoelastic relaxation region (Section 9.4).

One assumption which is made in studying liquids with ultrasonic longitudinal waves is that the absorption coefficients of different processes are additive. Truesdell (1953) has shown that this is only an approximation. At low temperatures in liquids where the absorption is high and a number of related processes are contributing to the absorption, it is possible that substantial errors arise in the determination of the structural relaxation region.

10.7 Sound Propagation in the Critical Region

The critical state of a one component fluid is that point on its (P, V, T) surface at which the specific volumes of the gas and liquid phases are identical. At the critical point both the heat capacity at constant pressure C_p and the isothermal compressibility β_T are infinite (Rowlinson, 1969). It has long been known that there is a sharp maximum in C_v at the critical point, and measurements of C_v of argon (Bagatskii, Voronel and Gusak, 1963), oxygen (Voronel and coworkers, 1964) and xenon (Edwards, Lipa and Buckingham, 1968) strongly suggest that C_v becomes infinite at the critical point.

The behaviour of the sound velocity in the critical region may be deduced by rewriting equation (10.5.3) using (10.5.4):

$$\frac{1}{\beta_S} - \frac{1}{\beta_T} = \frac{T(\partial P/\partial T)_V^2}{\rho C_v} \tag{10.7.1}$$

where $(\partial P/\partial T)_V$ is the thermal pressure coefficient which is related to the expansion coefficient θ and the isothermal compressibility β_T (Rowlinson, 1969):

$$\theta = \beta_T (\partial P/\partial T)_V \tag{10.7.2}$$

Hence

$$U^2 = 1/\rho\beta_S \tag{2.1.11}$$

$$= \frac{1}{\rho}\left[\frac{1}{\beta_T} + \frac{T(\partial P/\partial T)_V^2}{\rho C_v}\right] \tag{10.7.3}$$

while at the critical point

$$U_c^2 = \frac{T(\partial P/\partial T)_V^2}{\rho^2 C_v} \tag{10.7.4}$$

Since $(\partial P/\partial T)_V$ is finite at the critical point, U_c will have a sharp minimum at the C_v maximum there, and U_c will become zero if C_v becomes infinite.

Most studies of ultrasonic propagation in the critical region do not have sufficiently fine temperature or pressure resolution to establish whether the ultrasonic velocity becomes zero at the critical point (Schneider, 1951; Parbrook and Richardson, 1952; Tielsch and Tanneberger, 1954; Breazale, 1962; Nozdrev, 1965). A careful determination of the velocity of 1 MHz ultrasonic waves at the critical point of helium ($T_c = 5\cdot199$ K, $P_c = 2\cdot26$ atm) was made by Chase, Williamson and Tisza (1964). They used the phase comparison method (Section 3.7) to determine velocity to $\pm0\cdot01\%$, and controlled the temperature to $\pm10^{-4}$ K and the pressure to $\pm10^{-3}$ atm. Figure 10.7.1 shows their results along the isobar $2\cdot260\pm0\cdot001$ atm. The observed velocity decreases sharply as the critical temperature is approached from either side. Although the ultrasonic absorption becomes so large that velocity determinations could not be made within $\pm2\times10^{-3}$ K of T_c the decrease in velocity is so great in this region that a zero value of U_c does not appear unreasonable.

Figure 10.7.2 shows the maximum in the ultrasonic absorption at the critical point of xenon (Chynoweth and Schneider, 1952). This high absorption is accompanied by a velocity dispersion of some 6%. Similar minima in velocity and maxima in the absorption coefficient are found at the liquid–liquid critical point in a two component system (Brown and

Richardson, 1959; Nozdrev, 1965), near the second-order phase transition in helium (Chase, 1958), and at the transition from normal liquid to liquid crystal (Gabrielli and Verdini, 1955).

Although the liquid viscosity in the neighbourhood of the critical point typically increases to twice the value expected from an extrapolation from

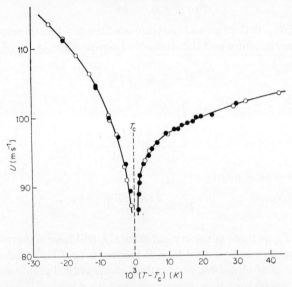

FIGURE 10.7.1. Ultrasonic velocity at 1 MHz as a function of temperature near the critical point of helium: pressure = $2 \cdot 260 \pm 0 \cdot 001$ atm; T_c, critical temperature; ○, measurements with temperature increasing; ●, measurements with temperature decreasing (after Chase, Williamson and Tisza, 1964).

lower temperatures, the observed ultrasonic absorption near the critical point is several powers of ten greater than the classical absorption calculated from equation (2.2.6) using the observed viscosity and a reasonable estimate of the thermal conductivity. This excess absorption has been attributed by Botch and Fixman (1965) to a coupling of the ultrasonic wave to the density fluctuations which exist in the liquid (cf. Sections 3.8 and 15.5). A typical correlation length for such a fluctuation might be 10^{-6} m which is much shorter than a typical sound wavelength of 10^{-3} m ($U = 1000$ m s^{-1}, $f = 1$ MHz). The fluctuations have a frequency spectrum, however, and there will be a transfer of energy from the sound wave to those fluctuations of high frequency and long wavelength which are in resonance with the sound wave. Because of the rapid equilibration of the fluctuations among themselves, the energy transfer from the sound wave is irreversible and sound absorption results. Botch and Fixman (1965) calculated the spectral density of the fluctuations and from this obtained a frequency dependent heat capacity of the liquid. Substitution of their values for C_v and C_p in equation (2.5.18) with $\omega\tau \ll 1$ give the sound absorption;

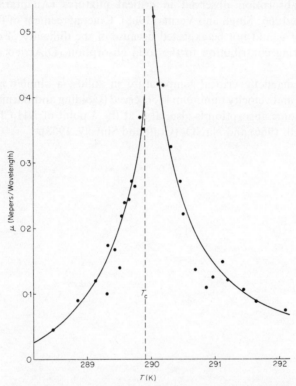

FIGURE 10.7.2. Ultrasonic absorption per wavelength at 250 kHz as a function of temperature near the critical point of xenon: T_c, critical temperature (after Chynoweth and Schneider, 1952).

the ultrasonic velocity was obtained from (10.5.5). This calculation gives excellent agreement with the observed absorption although there are discrepancies between the observed and calculated sound velocities. A similar treatment has been successful in calculating the ultrasonic absorption in critical mixtures (Fixman, 1962).

When the density fluctuations become comparable to the sound wavelength, scattering of the sound waves would be expected (cf. Section 15.3). Brown and Richardson (1959) have detected the energy scattered from acoustic pulses with a transducer probe near the critical point of an aniline–cyclohexane mixture. From the frequency dependence of the ultrasonic absorption in HCl, however, Breazale (1962) concludes that scattering plays only a small part in the ultrasonic absorption in the critical region. In general, there is reasonable agreement between Fixman's (1962) theory and the absorption observed in critical mixtures (Anantaraman and coworkers, 1966; Singh and Verma, 1968). Exact agreement of theory and experiment would not be expected because of the difficulty of estimating the scattering contribution to the total absorption (D'Arrigo and Sette, 1968).

At the magnetic critical temperature in solids, a similar absorption maximum and velocity minimum is observed (Golding and Barmatz, 1969). The ultrasonic absorption is also large at the λ point of NH_4Cl (Garland and Yarnell, 1966) and $NaNO_3$ (Craft and Slutsky, 1968).

CHAPTER 11

MOLECULAR ENERGY TRANSFER IN LIQUIDS

11.1 Introduction

When a longitudinal sound wave is passed through a liquid, the translational energy of the liquid molecules alters in phase with the sound wave. As in gases (Chapter 4), transfer of energy between the translational and the vibrational and rotational modes can occur as a result of molecular collisions. In a liquid the free volume is typically of the order of 10% of the specific volume, and the average distance between the centres of two molecules is only slightly greater than the collision diameter. A molecule experiences a strong repulsive force when it tries to pass between its neighbours, and the rate of diffusion of molecules in a liquid is much slower than in a gas. During the time between diffusional jumps, a molecule oscillates back and forth many times within a cage of its neighbours, the end of each oscillation occurring as a collision.

In associated liquids the translational motion of a molecule in its cage and the internal vibrational motion of the molecule are strongly coupled. Hence the translational–vibrational relaxation time will be very short ($\sim 10^{-12}$ s). In non-associated liquids, however, the coupling between the internal and external degrees of freedom is weaker, and the vibrational relaxation times are longer. Since there are approximately 10^3 times as many collisions per second in a liquid as in a gas at atmospheric pressure at the same temperature, the vibrational relaxation time of the liquid should be about 10^{-3} times that in the gas if the same factors govern translational–vibrational energy transfer in the two phases. Thus the vibrational relaxation times of most liquids will be too short to be observed in the ultrasonic range, although they are often accessible to hypersonic techniques.

The rotational motion of liquids is strongly coupled to the translational motion. As a result, rotational–translational relaxation times of liquids are too short to be observed by acoustic techniques in the available frequency range below 10 GHz.

11.2 Ultrasonic Studies of Vibrational Relaxation in Liquids

The first observations of vibrational relaxation in liquids were by Rapuano (1950) and by Lamb and Andreae (1951) in carbon disulphide. Subsequently Andreae, Heasell and Lamb (1956) carried out a detailed investigation of this liquid. They measured the ultrasonic absorption at frequencies up to 190 MHz at 210 K and 298 K, and their results at 298 K are shown in Figure 11.2.1. From the known values of the vibrational

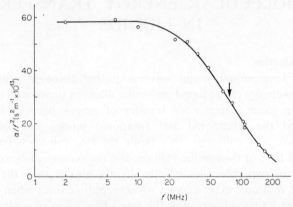

FIGURE 11.2.1. Ultrasonic absorption in liquid carbon disulphide at 298 K: —, calculated from equation (2.5.21): arrow indicates characteristic frequency (after Andreae, Heasell and Lamb, 1956).

frequencies of CS_2, the relaxational heat capacity may be calculated and hence the theoretical curve for a single relaxation process may be obtained from equation (2.5.21). There is excellent agreement between the observed values of the absorption coefficient and those calculated assuming the relaxation of the total vibrational heat capacity. The characteristic frequencies of the relaxation are 31 MHz at 210 K and 78 MHz at 298 K.

TABLE 11.2.1. Effect of pressure on the characteristic frequency of the relaxation process in liquid carbon disulphide at 244 K (Litovitz, Carnevale and Kendall, 1957)

P (atm)	3	443	720	978
f_c (MHz)	46·3	52·2	58·3	69·3

Litovitz, Carnevale and Kendall (1957) have investigated the effect of hydrostatic pressure on this relaxation, and their results are given in Table 11.2.1. The primary effect on a liquid of an increase in pressure is a decrease in the free volume, and this leads to an increase in the rate of molecular collisions. Hence a process which is dependent on the rate of molecular collisions should become more rapid as the pressure is increased, and this is observed in CS_2. It is also found that the characteristic frequency of the relaxation increases linearly with the concentration of added alcohol impurity (Slie and Litovitz, 1961). Both these effects provide evidence that the process being observed is vibrational relaxation.

In liquid carbon dioxide the relaxation region found by Bass and Lamb (1958) in the temperature range 273–323 K can also be described by a single relaxation time involving the whole of the vibrational heat capacity. CO_2 has also been studied by Henderson and Peselnick (1957) and Madigosky and Litovitz (1961) over a wide range of pressure from the gaseous state through the critical region into the liquid state. In all cases a single relaxation time describes the velocity dispersion and absorption, the vibrational relaxation time being inversely proportional to the density of the gas or liquid at all but the highest densities.

Bass and Lamb (1958) have also observed vibrational relaxation with the ultrasonic method in liquid sulphur hexafluoride, nitrous oxide, cyclopropane and methyl chloride. In these liquids the characteristic frequency of the relaxation fell above the highest frequency which was available experimentally (50 MHz), and so a complete analysis was not possible. Values of the vibrational relaxation times in Table 11.2.2 have been deduced on the assumption that the whole of the vibrational heat capacity relaxes with a single relaxation time.

Bass and Lamb (1957) measured the absorption of ultrasonic waves in liquid sulphur dioxide over the frequency and temperature range 3–45 MHz and 273–323 K. A pronounced relaxation region is found with a characteristic frequency of about 23 MHz (Figure 11.2.2), but this relaxation is incompatible with the relaxation of all three fundamental modes of vibration of SO_2. The vibrational mode which does not participate is that of lowest frequency, and this is expected to relax at a much higher frequency. The vibrational relaxation of liquid SO_2 is thus similar to that of the gas where two relaxation regions are also found (Section 4.5).

Two vibrational relaxation regions have also been found in liquid dichloromethane (Andreae, Joyce and Oliver, 1960; Hunter and Dardy, 1965) and dibromomethane (Hunter and Dardy, 1966). In each case the vibrational frequency of the lowest mode is substantially lower than the frequencies of the other modes and does not take part in the lower frequency relaxation (cf. Section 4.5).

Table 11.2.2 lists the vibrational relaxation times of some liquids determined by the ultrasonic technique. A decrease in α/f^2 between 1 and 2 GHz in chloroform, thiophene, fluorobenzene, chlorobenzene, bromobenzene and toluene also indicates the onset of vibrational relaxation (Parpiev, Khabibullaev and Khaliulin, 1970). In most liquids, however, vibrational relaxation occurs at frequencies above the ultrasonic range.

TABLE 11.2.2. Vibrational relaxation times in liquids determined from studies of ultrasonic propagation

Liquid	T (K)	Pressure (atm)	τ (ns)	Reference*
Cl_2	273	s[a]	10·3	1
	303	s	8·3	1
N_2O	298	s	2·6	2
CO_2	273	70	15·0	2
	324	7·3	760	3
	324	225	15·1	3
$SO_2{}^b$	298	10	6·8	4
CS_2	298	s	2·0	5
CH_4	244	360	3·6	6
	303	525	2·1	6
CH_3Cl	298	s	0·65	2
$CH_2Cl_2{}^b$	213	s	1·4	7
	298	s	1·0	7
CCl_4	300	s	0·090	8
$CH_2Br_2{}^b$	233	s	0·58	9
	298	s	0·42	9
$SiCl_4$	300	s	0·053	8
$GeCl_4$	300	s	0·055	8
$SnCl_4$	300	s	0·057	8
$TiCl_4$	300	s	0·032	8
SF_6	298	s	2·5	2
Cyclo-C_3H_6	298	s	0·90	2
C_5H_5N	298	s	0·19	10
C_6H_6	298	s	0·26	11

[a] s = saturation vapour pressure.
[b] Longest relaxation time only.

* References

1. Sittig (1960).
2. Bass and Lamb (1958).
3. Henderson and Peselnick (1957).
4. Bass and Lamb (1957).
5. Andreae, Heasell and Lamb (1956).
6. Madigosky (1963).
7. Hunter and Dardy (1965).
8. Rasmussen (1968).
9. Hunter and Dardy (1966).
10. Parpiev and coworkers (1970).
11. Khabibullaev and Khaliulin (1969).

11.3 Hypersonic Studies of Vibrational Relaxation in Liquids

The conventional range of ultrasonic frequencies may conveniently be extended to 10^{10} Hz by the use of Brillouin scattering (Section 3.8). Hitherto most workers have studied the dispersion of the hypersonic velocity, and techniques are also available for studying hypersonic absorption. Table 11.3.1 contains values of the ultrasonic and hypersonic velocities in some typical simple liquids. There is a pronounced velocity dispersion in the non-associated liquids and this may be attributed to

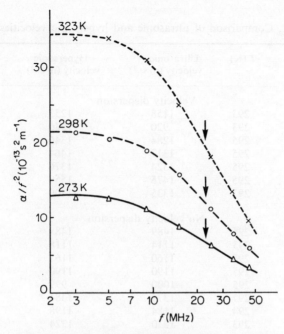

FIGURE 11.2.2. Ultrasonic absorption in liquid
sulphur dioxide: —, △, 273 K; ---, ○, 298 K;
---, ×, 323 K: lines are calculated from equation
(2.5.21): arrows indicate characteristic frequencies
(after Bass and Lamb, 1957).

vibrational relaxation. In liquids which are associated or composed of very
flexible molecules, there is no evidence of velocity dispersion. This confirms
the close coupling between translational motion and the internal energy in
such liquids.

By measuring the hypersonic velocity as a function of scattering angle,
the hypersonic velocity may be measured as a function of frequency. Hence
the vibrational relaxation time is obtained (Chapter 2), and some results
are given in Table 11.3.2. Where comparison is possible, there is good
agreement between these values and the ultrasonic results in Table 11.2.2.

Hypersonic velocity measurements have also been used to confirm that
two vibrational regions occur in liquid dichloromethane (Clark and
coworkers, 1966). Figure 11.3.1 shows the observed temperature dependence of the velocity of sound waves of frequencies ~ 5 GHz and 30 MHz:
the latter frequency is too low for vibrational relaxation to occur over the
range of temperature studied. Also shown are the velocities calculated on
the assumption that there is no vibrational relaxation, that modes 2–9 are

TABLE 11.3.1. Comparison of ultrasonic and hypersonic velocities in liquids

Liquid	T (K)	Ultrasonic velocity (m s^{-1})	Hypersonic velocity (m s^{-1})	Reference*
		Velocity dispersion		
CS_2	293	1158	1253	1
CCl_4	293	920	1040	2
Cyclo-C_6H_{12}	295	1284	1346	3
C_6H_6	295	1317	1464	4
$C_6H_5CH_3$	295	1324	1376	3
$C_6H_5NO_2$	295	1475	1559	3
C_6H_5COCl	297	1335	1400	5
		No velocity dispersion		
H_2O	295	1489	1488	3
CH_3OH	295	1114	1118	3
C_2H_5OH	295	1160	1162	3
CH_3COCH_3	295	1190	1190	3
$C_2H_5OC_2H_5$	295	1000	999	3
1,4-Dioxan	297	1379	1384	5
n-C_8H_{16}	294	1184	1198	6
n-C_8H_{18}	293	1230	1224	6

* References
1. Chiao and Stoicheff (1964).   4. Benedek and coworkers (1964).
2. Fabelinskii (1957).   5. Hakim and Comley (1965).
3. Fleury and Chiao (1966).   6. Comley (1966).

TABLE 11.3.2. Vibrational relaxation times in liquids from measurements of hypersonic velocity

Liquid	T (K)	τ (ns)	Reference*
CS_2	293	2·1	1
CCl_4	293	0·10	1
C_6H_6	293	0·27	2
Cyclo-C_6H_{12}	295	0·08	3
$C_6H_5CH_3$	295	0·006	3
$C_6H_5NO_2$	295	0·009	3

* References   2. Caloin and Candau (1969).
1. Barocchi, Mancini and Vallauri (1968).   3. Fleury and Chiao (1966).

relaxing, and that all the vibrational modes are relaxing. At high temperatures only the lowest frequency mode of vibration is able to equilibrate with the hypersonic wave. At lower temperatures vibrational energy

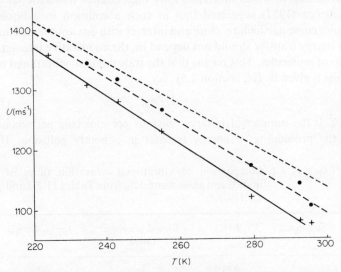

FIGURE 11.3.1. Ultrasonic velocity at 30 MHz (+) and hypersonic velocity (●) in liquid dichloromethane: —, sound velocity with all vibrational modes contributing to the specific heat; ―――, sound velocity with no contribution from vibrational modes 2–9; ----, sound velocity with no contribution from vibrational modes (after Clark and coworkers, 1966).

transfer occurs more slowly, and the lowest vibrational mode also shows vibrational relaxation.

11.4 Theories of Vibrational Relaxation in Liquids

There have been two main approaches to the problem of calculating vibrational relaxation times in liquids. In the isolated binary collision theory it is assumed that the probability of energy transfer in a collision is the same in the liquid as in the gas, and that the ratio of the vibrational relaxation times in the two phases equals the ratio of the number of binary collisions. In the cooperative interaction theory it is assumed that vibrational relaxation occurs as a result of the cooperative motion of several molecules within an encounter cage.

The original derivation of the isolated binary collision theory was given by Litovitz (1957). He assumed that the same mechanism is responsible for vibrational–translational energy transfer in the liquid as in the gas. In general there is little change in the vibrational frequencies or in the translational velocities of unassociated molecules in going from the gaseous to the liquid phase. The main contribution to energy transfer comes from those collisions in which molecules have high relative translational velocities. Litovitz (1957) suggested that in such a collision in a liquid two molecules come particularly close and interact with one another exclusively. Hence energy transfer should not depend on the cooperative interaction of a group of molecules. This means that the translational–vibrational relaxation time is given by (cf. Section 4.5)

$$\tau = 1/Zp_{10} \tag{11.4.1}$$

where Z is the number of binary collisions per molecule per second and p_{10} is the probability of energy transfer in a binary collision. If Z is

TABLE 11.4.1. Comparison of vibrational relaxation times of liquids and gases using data from Tables 11.2.2 and 4.5.1

Substance	T (K)	Liquid pressure (atm)	$(\rho\tau)_{liq}/(\rho\tau)_{gas}$
Cl_2	273	s[a]	0·86
	303	s	0·99
N_2O	298	s	1·1
CO_2	298	69	1·0
	324	225	1·1
SO_2[b]	298	10	6·4
CS_2	298	s	1·6
CH_3Cl	298	s	1·6
CH_2Cl_2[b]	298	s	4·4
CCl_4	300	s	1·1
$SiCl_4$	300	s	1·8
$GeCl_4$	300	s	3·9
$SnCl_4$	300	s	5·2
$TiCl_4$	300	s	>3
SF_6	298	s	0·79
Cyclo-C_3H_6	298	s	1·3

[a] s = saturation vapour pressure.
[b] Longest relaxation time only.

proportional to the number of molecules in unit volume, the product $\rho\tau$ should be constant in both gas and liquid phases, ρ being the density.

Table 11.4.1 shows that the ratio $(\rho\tau)_{liq}/(\rho\tau)_{gas}$ is close to unity in many liquids, while in Figure 11.4.1 the ratio of the characteristic frequency f_c to ρ is constant except at high densities in CO_2. This suggests that the isolated binary collision theory is a reasonable approximation in liquids at moderate densities.

In calculating τ from equation (11.4.1), p_{10} is taken to be the same in the liquid as in the gas phase. Z is given by the ratio of the mean molecular

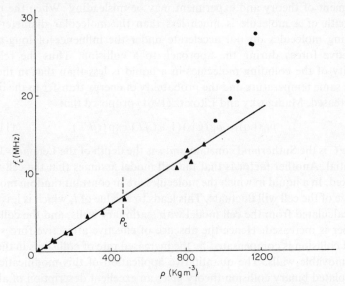

FIGURE 11.4.1. Variation of relaxation frequency f_c with density in gaseous and liquid carbon dioxide at 323·4 K: ρ_c, critical density; —, prediction of kinetic theory $f_c/\rho =$ constant; ▲, Henderson and Peselnick (1957); ●, Madigosky and Litovitz, 1961 (after Madigosky and Litovitz, 1961).

translational velocity c to the mean free path l_f of the molecules in the liquid:

$$Z = c/l_f \tag{11.4.2}$$

Although gas kinetic theory can be used to calculate c, it is more difficult to estimate l_f in a liquid. Litovitz (1957) examined a number of methods of calculating l_f, and concluded that the Eyring and Hirschfelder (1937) cell model provided a satisfactory estimate. The cell model assumes that the mean free path depends on the difference between the average cell diameter and the molecular diameter σ:

$$l_f = 2[(1/\rho)^{\frac{1}{3}} - \sigma] \tag{11.4.3}$$

Hence

$$\tau = 2[(1/\rho)^{\frac{1}{3}} - \sigma]/cp_{10} \tag{11.4.4}$$

The combination of the cell model with the isolated binary collision theory gives an excellent description of both the pressure and temperature dependence of the vibrational relaxation time of CO_2 over the entire range of density.

Madigosky and Litovitz (1961) have suggested, however, that this good agreement of theory and experiment may be misleading. When the mean free path of a molecule is much less than the molecular diameter, the colliding molecules do not accelerate under the influence of long-range attractive forces during the approach to a collision. Thus the relative velocity of the colliding molecules in a liquid is less than that in the gas at the same temperature and the probability of energy transfer in the liquid is decreased. Madigosky and Litovitz (1961) proposed that

$$p_{10}(\text{liq}) = p_{10}(\text{gas})(1 + C/T)/\exp(\varepsilon/kT) \qquad (11.4.5)$$

where C is the Sutherland constant and ε is the depth of the Lennard–Jones potential. Another factor is that the cell model assumes that the cell walls are fixed. In a liquid in which the molecules are in constant random motion, the size of the cell will fluctuate. This leads to a value of l_f which is less than that calculated from the cell model with stationary walls, and the collision number is increased. Hence the absence of effective attractive forces in a liquid collision is compensated by the increased rate of collisions in the cell with movable walls. The quantitative application of this modification of the isolated binary collision theory gives an excellent description of all the results available for CO_2 and CS_2 (Madigosky and Litovitz, 1961), CH_4 (Madigosky, 1963) and the Group IV tetrahalides (Rasmussen, 1968).

In contrast to the isolated binary collision theory, cooperative interaction theories have proved much less successful. One of the earliest calculations was that of Herzfeld (1952, 1957) who considered a molecule vibrating in the centre of a spherical cell, the walls of which are made up of nearest neighbours. He assumed that the probability of energy transfer is determined by the cooperative interaction of neighbouring molecules having the average kinetic energy, and he used the Lennard–Jones potential to calculate the coupling between inter- and intramolecular vibrations. This treatment resulted in a calculated value of the vibrational relaxation time some hundred times greater than that observed.

Fixman (1961) and Zwanzig (1961) have considered situations intermediate between the isolated binary collision theory and the full cooperative interaction theory. Fixman (1961) maintained the separation of the relaxation process into the product of the probability of energy transfer in a collision and the number of collisions. He found that the random Brownian motion of molecules in the liquid makes low energy collisions much more effective in promoting energy transfer in liquids than in gases. As the liquid density decreases, the perturbing effect of neighbouring molecules also decreases so that the isolated binary collision theory becomes applicable.

Zwanzig (1961) showed that the probability of energy transfer is proportional to the spectral density of the force exerted on the oscillator by its surroundings. The isolated binary collision theory is found to apply when the collision frequency is much smaller than the vibrational frequency as in a low density fluid, but in liquids where the collision frequency is comparable to the vibration frequency it is inapplicable. The latter conclusion has been challenged by Herzfeld (1962), however. He suggests that since very few collisions are effective in leading to energy transfer, the condition that the effective collisional frequency be less than the vibrational frequency holds, and the isolated binary collision theory remains valid.

It is clear that the isolated binary collision theory plus certain assumptions on liquid structure is adequate to describe the pressure and temperature dependence of the vibrational relaxation time in gases and liquids up to moderate densities. Because the relaxation times are so short, however, little data is available for high density liquids or for liquids of more complex molecular structure. In these cases the criticisms of the isolated binary collision theory may be valid and cooperative effects may become important. This problem will not be settled until experimental techniques become available for measuring vibrational relaxation times of 10^{-10} s and below.

CHAPTER 12

ROTATIONAL ISOMERIC RELAXATION IN LIQUIDS

12.1 Introduction

Whenever molecules can occupy two or more internal energy levels the equilibrium distribution of molecules among these levels will be perturbed by the passage of a sound wave. At sufficiently low sound frequencies, the relatively long interval between successive cycles allows plenty of time for energy to be transferred from the sound wave to the internal degrees of freedom. As the sound frequency increases, a time-lag occurs between the temperature and pressure variations in the sound wave and the adjustment of the internal equilibrium, and this leads to a maximum in the ultrasonic absorption per wavelength and a decrease in α/f^2 (Section 2.5).

Figure 12.1.1 shows the energy level diagram for a two-state equilibrium (cf. Section 2.3)

$$A \underset{k_b}{\overset{k_f}{\rightleftharpoons}} B \qquad (12.1.1)$$

in which B is the higher energy state. Studies of liquids with ultrasonic longitudinal waves provide information on the rate constants for the

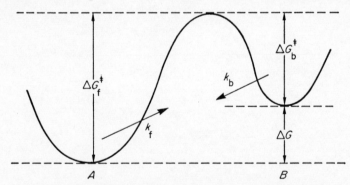

FIGURE 12.1.1. Energy level diagram for a two-state equilibrium.

forward and backward reactions, k_f and k_b, the free energy barrier $\Delta G_b^{\neq}$, and the free energy difference between the two levels, ΔG. These quantities are calculated from the variation of the absorption per wavelength μ with frequency and temperature. At a given temperature μ has a maximum value μ_m at a characteristic frequency f_c (Section 2.5): from the temperature dependence of f_c, $\Delta G_b^{\neq}$ may be found, while the temperature dependence of μ_m yields ΔG. The theoretical basis of this calculation is outlined in the following Sections, and some representative experimental results are discussed.

12.2 Determination of Energy Barriers

In the two-state equilibrium (12.1.1), we assume that each reaction is unimolecular and takes place in an ideal solution. The characteristic frequency f_c of the equilibrium is related to the rate constants by the expression (Section 2.3)

$$f_c = \frac{1}{2\pi\tau} = \frac{k_f + k_b}{2\pi} \tag{12.2.1}$$

where τ is the relaxation time of the equilibrium. The theory of rate processes (Glasstone, Laidler and Eyring, 1941) expresses the rate constant k_r of a reaction in terms of the increase in free energy $\Delta G^{\neq}$ on moving from the initial state to an intermediate activated state:

$$k_r = \frac{kT}{h}\exp(-\Delta G^{\neq}/RT) \tag{12.2.2}$$

Hence

$$f_c = \frac{1}{2\pi}\frac{kT}{h}[\exp(-\Delta G_f^{\neq}/RT) + \exp(-\Delta G_b^{\neq}/RT)] \tag{12.2.3}$$

If we assume that equilibrium in this reaction lies well to the left, that is $\Delta G_b^{\neq} \ll \Delta G_f^{\neq}$,

$$f_c = \frac{1}{2\pi}\frac{kT}{h}\exp(-\Delta G_b^{\neq}/RT)$$

$$= \frac{1}{2\pi}\frac{kT}{h}\exp(\Delta S_b^{\neq}/R)\exp(-\Delta H_b^{\neq}/RT) \tag{12.2.4}$$

Experimentally f_c is measured over as wide a range of temperature as possible, and $\Delta H_b^{\neq}$ and $\Delta S_b^{\neq}$ determined from the slope and intercept of a graph of $\log(f_c/T)$ against $1000/T$. When $\Delta G_b^{\neq} \ll \Delta G_f^{\neq}$, $k_b \gg k_f$, and k_b is obtained from equation (12.2.1):

$$k_b \approx 2\pi f_c \tag{12.2.5}$$

12.3 Determination of the Energy Difference between Two States

In this Section we derive a relationship between the absorption per wavelength μ and the adiabatic compressibility β_S of a liquid. β_S is then expressed in terms of the heat capacity of the liquid, and from the latter the free energy difference between two states may be calculated.

In Chapter 10, equation (10.3.12) related the ultrasonic absorption to the elastic constants of the liquid:

$$\mu = \frac{\pi}{\rho U^2}(K'' + \tfrac{4}{3}G'') \tag{10.3.12}$$

It is generally found that the observed values of the absorption are sufficiently small for the approximations involved in the derivation of this equation to be valid. Most acoustic studies of rotational isomerization are carried out in Newtonian liquids in which the loss modulus G'' is given by

$$G'' = \omega \eta_s \tag{5.3.18}$$

If the ultrasonic absorption μ' arising from a two-state equilibrium is additive with that arising from other sources,

$$\mu' = \frac{\pi}{\rho U^2}K'' = \frac{\pi}{\rho U^2}K_2\frac{\omega\tau}{1+\omega^2\tau^2} \tag{12.3.1}$$

using equation (10.2.6). K_2 is the relaxational part of the bulk modulus associated with the equilibrium, while τ is the relaxation time. We now write

$$K^* = 1/\beta_S^* \tag{12.3.2}$$

where β_S^* is the complex adiabatic compressibility of the liquid, and

$$K_0 = 1/\beta_S; \quad K_\infty = 1/\beta_{S\infty} \tag{12.3.3}$$

Hence

$$K_2 = K_\infty - K_0$$

$$= (1/\beta_{S\infty}) - (1/\beta_S)$$

$$= \beta_{Sr}/\beta_S\beta_{S\infty} \tag{12.3.4}$$

where the relaxational adiabatic compressibility is

$$\beta_{Sr} = \beta_S - \beta_{S\infty} \tag{12.3.5}$$

Equation (12.3.1) may thus be written

$$\mu' = \frac{\pi}{\rho U^2}\frac{\beta_{Sr}}{\beta_S\beta_{S\infty}}\frac{\omega\tau}{1+\omega^2\tau^2} \tag{12.3.6}$$

The velocity dispersion in rotational isomerization is generally small. Hence (cf. 2.1.11)

$$\rho U^2 \approx \rho U_\infty^2 = 1/\beta_{S\infty} \tag{12.3.7}$$

and (12.3.6) reduces to

$$\mu' = \pi \frac{\beta_{Sr}}{\beta_S} \frac{\omega\tau}{1+\omega^2\tau^2} \tag{12.3.8}$$

This equation relates the ultrasonic absorption to the relaxational adiabatic compressibility and the relaxation time of a two-state equilibrium. It is not exact since it is based on the assumptions that the velocity dispersion is small and that equation (10.3.12) is applicable.

The passage of a longitudinal sound wave through a liquid causes small variations about the equilibrium values of the pressure P, volume V, temperature T and entropy S at the sound frequency $\omega/2\pi$. Following Andreae and Lamb (1956) we assume that these variations are governed by an equation of state

$$X = F_\omega(Y,Z) \tag{12.3.9}$$

where X, Y and Z are any three of the variables P, V, T and S, and F_ω is a function depending on ω. If Y and Z vary sinusoidally it is assumed that the departures from equilibrium are so small that X will respond linearly and sinusoidally. Hence

$$dX = \left(\frac{\partial X}{\partial Y}\right)_{Z,\omega} dY + \left(\frac{\partial X}{\partial Z}\right)_{Y,\omega} dZ \tag{12.3.10}$$

We also assume that the effect of a variation of frequency on the equation of state (12.3.9) arises only from the existence of two different internal energy states in the molecule:

$$X = F(Y,Z,y) \tag{12.3.11}$$

Here y is a reaction variable describing the position of the equilibrium between the two energy states. For small displacements from equilibrium we have

$$dX = \left(\frac{\partial X}{\partial Y}\right)_{Z,y} dY + \left(\frac{\partial X}{\partial Z}\right)_{Y,y} dZ + \left(\frac{\partial X}{\partial y}\right)_{Y,Z} dy \tag{12.3.12}$$

Combination of (12.3.10) and (12.3.12) gives

$$\left(\frac{\partial X}{\partial Z}\right)_{Y,\omega} = \left(\frac{\partial X}{\partial Z}\right)_{Y,y} + \left(\frac{\partial X}{\partial y}\right)_{Y,Z}\left(\frac{\partial y}{\partial Z}\right)_{Y,\omega} \tag{12.3.13}$$

The term $(\partial X/\partial Z)_{Y,y}$ in equation (12.3.13) is independent of the frequency of the applied sound wave since it contains no contribution from the internal energy states. The second term is a relaxational term, and represents the contribution of the internal energy states to $(\partial X/\partial Z)_{Y,\omega}$.

Equation (12.3.13) can now be used to expand: (a) the adiabatic compressibility β_S; (b) the isothermal compressibility β_T; (c) the thermal expansion coefficient θ; and (d) the specific heat at constant pressure C_p in terms of the extent of reaction y.

(a) $\quad \beta_S = -\dfrac{1}{V}\left(\dfrac{\partial V}{\partial P}\right)_{S,\omega} = -\dfrac{1}{V}\left(\dfrac{\partial V}{\partial P}\right)_{S,y} - \dfrac{1}{V}\left(\dfrac{\partial V}{\partial y}\right)_{S,P}\left(\dfrac{\partial y}{\partial P}\right)_{S,\omega}$

$$= \beta_{S\infty} + \beta_{Sr} \qquad (12.3.14)$$

Hence

$$\beta_{Sr} = -\dfrac{1}{V}\left(\dfrac{\partial V}{\partial y}\right)_{S,P}\left(\dfrac{\partial y}{\partial P}\right)_{S,\omega} \qquad (12.3.15)$$

(b) $\quad \beta_T = -\dfrac{1}{V}\left(\dfrac{\partial V}{\partial P}\right)_{T,\omega} = -\dfrac{1}{V}\left(\dfrac{\partial V}{\partial P}\right)_{T,y} - \dfrac{1}{V}\left(\dfrac{\partial V}{\partial y}\right)_{T,P}\left(\dfrac{\partial y}{\partial P}\right)_{T,\omega}$

$$= \beta_{T\infty} + \beta_{Tr} \qquad (12.3.16)$$

Here $\beta_{T\infty}$ is the value of β_T when $\omega\tau \gg 1$ and β_{Tr} is the relaxational isothermal compressibility.
Hence

$$\beta_{Tr} = -\dfrac{1}{V}\left(\dfrac{\partial V}{\partial y}\right)_{T,P}\left(\dfrac{\partial y}{\partial P}\right)_{T,\omega} = -\dfrac{\Delta V}{V}\left(\dfrac{\partial y}{\partial P}\right)_{T,\omega} \qquad (12.3.17)$$

where the volume change as a result of the reaction for unit change in y is

$$\Delta V = \left(\dfrac{\partial V}{\partial y}\right)_{T,P} \qquad (12.3.18)$$

(c) $\quad C_p = T\left(\dfrac{\partial S}{\partial T}\right)_{P,\omega} = T\left(\dfrac{\partial S}{\partial T}\right)_{P,y} + T\left(\dfrac{\partial S}{\partial y}\right)_{P,T}\left(\dfrac{\partial y}{\partial T}\right)_{P,\omega}$

$$= C_{p\infty} + C_i \qquad (12.3.19)$$

where C_i is the relaxational heat capacity. Hence

$$C_i = T\left(\dfrac{\partial S}{\partial y}\right)_{P,T}\left(\dfrac{\partial y}{\partial T}\right)_{P,\omega} = \Delta H\left(\dfrac{\partial y}{\partial T}\right)_{P,\omega} \qquad (12.3.20)$$

where the enthalpy change as a result of the reaction for unit change in y is

$$\Delta H = T\left(\dfrac{\partial S}{\partial y}\right)_{P,T} \qquad (12.3.21)$$

(d) $\quad \theta = \dfrac{1}{V}\left(\dfrac{\partial V}{\partial T}\right)_{P,\omega} = \dfrac{1}{V}\left(\dfrac{\partial V}{\partial T}\right)_{P,y} + \dfrac{1}{V}\left(\dfrac{\partial V}{\partial y}\right)_{P,T}\left(\dfrac{\partial y}{\partial T}\right)_{P,\omega}$

$$= \theta_\infty + \theta_r \qquad (12.3.22)$$

Hence

$$\theta_r = \frac{1}{V}\left(\frac{\partial V}{\partial y}\right)_{P,T}\left(\frac{\partial y}{\partial T}\right)_{P,\omega} = \frac{\Delta V}{V}\left(\frac{\partial y}{\partial T}\right)_{P,\omega} \tag{12.3.23}$$

by (12.3.18). An alternative expression for θ_r can be obtained using Maxwell's relation

$$\left(\frac{\partial V}{\partial T}\right)_P = -\left(\frac{\partial S}{\partial P}\right)_T \tag{12.3.24}$$

Hence

$$\theta = -\frac{1}{V}\left(\frac{\partial S}{\partial P}\right)_{T,\omega} = -\frac{1}{V}\left(\frac{\partial S}{\partial P}\right)_{T,y} -\frac{1}{V}\left(\frac{\partial S}{\partial y}\right)_{P,T}\left(\frac{\partial y}{\partial P}\right)_{T,\omega}$$

$$= \theta_\infty + \theta_r \tag{12.3.25}$$

and

$$\theta_r = -\frac{1}{V}\left(\frac{\partial S}{\partial y}\right)_{P,T}\left(\frac{\partial y}{\partial P}\right)_{T,\omega} = -\frac{\Delta H}{VT}\left(\frac{\partial y}{\partial P}\right)_{T,\omega} \tag{12.3.26}$$

by (12.3.21). It thus follows from equations (12.3.20) and (12.3.23) that

$$C_i = V\frac{\Delta H}{\Delta V}\theta_r = \left(\frac{\Delta H}{\Delta V}\right)^2\frac{V}{T}\beta_{Tr} \tag{12.3.27}$$

from (12.3.17) and (12.3.26).

We can now relate β_{Sr} to C_p using the expansion

$$\beta_{Sr} = -\frac{1}{V}\left(\frac{\partial V}{\partial y}\right)_{S,P}\left(\frac{\partial y}{\partial P}\right)_{S,\omega} \tag{12.3.15}$$

$$= -\frac{1}{V}\left[\left(\frac{\partial V}{\partial y}\right)_{P,T} - \left(\frac{\partial V}{\partial S}\right)_{P,y}\left(\frac{\partial S}{\partial y}\right)_{P,T}\right]\left[\left(\frac{\partial y}{\partial P}\right)_{T,\omega} + \left(\frac{\partial y}{\partial T}\right)_{P,\omega}\left(\frac{\partial T}{\partial P}\right)_{S,\omega}\right] \tag{12.3.28}$$

using (12.3.13). Equation (12.3.28) may be simplified by using the relation

$$\left(\frac{\partial T}{\partial P}\right)_S = \frac{T}{C_p}\left(\frac{\partial V}{\partial T}\right)_P \tag{12.3.29}$$

along with equations (12.3.18), (12.3.19), (12.3.21), (12.3.22), (12.3.23), (12.3.26), (12.3.27) and (12.3.29) to give

$$\beta_{Sr} = VTC_i\left(\frac{\theta_\infty}{C_{p\infty}} - \frac{\Delta V}{V\Delta H}\right)\left(\frac{\theta}{C_p} - \frac{\Delta V}{V\Delta H}\right)$$

$$= \frac{VTC_i}{C_{p\infty}C_p}\left(\theta_\infty - \frac{\Delta VC_{p\infty}}{V\Delta H}\right)\left(\theta - \frac{\Delta VC_p}{V\Delta H}\right) \tag{12.3.30}$$

Now

$$\theta_\infty - \frac{\Delta V C_{p\infty}}{V\Delta H} = (\theta - \theta_r) - \frac{\theta_r}{C_i}(C_p - C_i)$$

$$= \theta - \frac{\Delta V C_p}{V\Delta H} \tag{12.3.31}$$

by (12.3.27). Hence

$$\beta_{Sr} = \frac{VTC_i}{C_{p\infty}C_p}\left(\theta - \frac{\Delta V C_p}{V\Delta H}\right)^2 \tag{12.3.32}$$

The thermodynamic relations

$$\beta_T - \beta_S = T\theta^2/\rho C_p \tag{12.3.33}$$

and

$$\beta_T/\beta_S = C_p/C_v \tag{12.3.34}$$

may be used to rearrange (12.3.32) to give

$$\frac{\beta_{Sr}}{\beta_S} = \frac{(\gamma-1)C_i}{C_{p\infty}}\left(1 - \frac{\Delta V C_p}{\Delta H V \theta}\right)^2 \tag{12.3.35}$$

Equation (12.3.35) can now be inserted into equation (12.3.8) to give

$$\mu' = \pi(\gamma-1)\frac{C_i}{C_{p\infty}}\left(1 - \frac{\Delta V C_p}{\Delta H V \theta}\right)^2 \frac{\omega\tau}{1+\omega^2\tau^2} \tag{12.3.36}$$

This equation with $\Delta V = 0$ has the same form as equation (2.5.18). It may also be rearranged to give an expression for α/f^2 identical to equation (2.5.21).

Equation (12.3.36) reaches a maximum value when

$$\omega_c\tau = 2\pi f_c\tau = 1 \tag{12.3.37}$$

The maximum value of μ' is then

$$\mu'_m = \frac{\pi(\gamma-1)}{2}\frac{C_i}{C_{p\infty}}\left(1 - \frac{\Delta V C_p}{\Delta H V \theta}\right)^2 \tag{12.3.38}$$

We now require to relate C_i to the extent of reaction y. For the two-state equilibrium between A and B in equation (12.1.1), we define y as

$$y = n_B/N \tag{12.3.39}$$

where n_B is the number of molecules of B, and

$$N = n_A + n_B \tag{12.3.40}$$

n_A being the number of A molecules. The rate of formation of B molecules is

$$\frac{dn_B}{dt} = k_f n_A - k_b n_B \tag{12.3.41}$$

This equation may be rearranged using (12.3.39) and

$$1 - y = n_A/N \tag{12.3.42}$$

to give

$$-\frac{dy}{dt} = k_b[y - K(1-y)]$$

$$= k_b(1+K)\left(y - \frac{K}{1+K}\right) \tag{12.3.43}$$

where $K = k_f/k_b$ is the equilibrium constant.

An alternative expression for dy/dt is given by equation (2.4.2):

$$-\frac{dy}{dt} = \frac{1}{\tau}(y - y_0) \tag{12.3.44}$$

where y_0 is the instantaneous equilibrium value of y. A comparison of (12.3.43) and (12.3.44) gives

$$y_0 = \frac{K}{1+K} \tag{12.3.45}$$

K is related to the free energy difference ΔG between the states

$$K = \exp(-\Delta G/RT) \tag{12.3.46}$$

Hence (12.3.45) can be rewritten

$$dy_0 = \frac{K d \ln(K)}{(1+K)^2} \tag{12.3.47}$$

We can now relate C_i to y using the expression for the frequency dependence of the relaxing heat capacity

$$C_i^\omega = T\left(\frac{\partial S}{\partial y}\right)_{P,T}\left(\frac{\partial y}{\partial T}\right)_{P,\omega} \tag{12.3.20}$$

$$= T\left(\frac{\partial S}{\partial y}\right)_{P,T}\left(\frac{\partial y_0}{\partial T}\right)_P\left(\frac{1}{1+i\omega\tau}\right) \tag{12.3.48}$$

using (2.4.4). Moreover,

$$C_i^\omega = \frac{C_i}{1+i\omega\tau} \tag{2.4.5}$$

Hence the total relaxational heat capacity is

$$C_i = T \left(\frac{\partial S}{\partial y}\right)_{P,T} \left(\frac{\partial y_0}{\partial T}\right)_P \qquad (12.3.49)$$

$$= \Delta H \frac{K}{(1+K)^2} \left(\frac{\partial \ln K}{\partial T}\right)_P \qquad (12.3.50)$$

using equations (12.3.21) and (12.3.47). By the use of the van't Hoff isochore, (12.3.50) may be rewritten

$$C_i = R \left(\frac{\Delta H}{RT}\right)^2 \frac{K}{(1+K)^2} \qquad (12.3.51)$$

and with (12.3.46) this becomes

$$C_i = R \left(\frac{\Delta H}{RT}\right)^2 \frac{\exp(-\Delta G/RT)}{[1+\exp(-\Delta G/RT)]^2} \qquad (12.3.52)$$

Hence

$$\mu'_m = \frac{\pi(\gamma-1)R}{2C_p} \left(1 - \frac{\Delta V C_p}{\Delta H V \theta}\right)^2 \left(\frac{\Delta H}{RT}\right)^2 \frac{\exp(-\Delta G/RT)}{[1+\exp(-\Delta G/RT)]^2} \qquad (12.3.53)$$

If $\Delta H = 0$, then C_i and μ'_m are also zero and no relaxational absorption occurs.

Equation (12.3.53) is an expression for the maximum value of the relaxational sound absorption per wavelength in terms of three unknown reaction parameters ΔV, ΔH and ΔG, each of which may vary with temperature. If experimental data for μ'_m were available over a range of temperature and pressure, then it would be possible to use the relations

$$\partial(\Delta G/T) \, \partial T = -\Delta H/T^2 \qquad (12.3.54)$$

and

$$\partial \Delta G/\partial P = \Delta V \qquad (12.3.55)$$

These equations can be integrated if it is assumed that ΔH and ΔV are independent of temperature. As this produces two constants of integration, unique values of the three reaction parameters cannot be obtained from equations (12.3.53), (12.3.54) and (12.3.55). Thus it is necessary to make various assumptions about the nature of the reaction. The significance of these assumptions has been investigated by Davies and Lamb (1959).

Generally there is no experimental information available over a range of pressure, and in order to estimate the energy differences it is usual to assume that ΔV is small and that $\Delta V C_p \ll \Delta H V \theta$. Equation (12.3.53) can then be written

$$\mu'_m = \frac{\pi(\gamma-1)R}{2C_p} \left(\frac{\Delta H}{RT}\right)^2 \frac{\exp(-\Delta G/RT)}{[1+\exp(-\Delta G/RT)]^2} \qquad (12.3.56)$$

The combination of equations (12.3.54) and (12.3.56) does not yield unique values for ΔH and ΔG because of the integration constant from equation (12.3.54). Hence the additional assumption is required that $\Delta G > 3RT$. Equation (12.3.56) can then be simplified with an error of less than 5%:

$$\mu'_m = \frac{\pi(\gamma-1)R}{2C_p}\left(\frac{\Delta H}{RT}\right)^2 \exp(-\Delta H/RT)\exp(\Delta S/R) \qquad (12.3.57)$$

ΔH can now be evaluated from the slope of a plot of $\log[C_p T^2 \mu'_m/(\gamma-1)]$ versus $1000/T$, and substitution of this value in equation (12.3.57) yields ΔS. In practice the values of C_p and γ may not be known with sufficient accuracy, and it is preferable to replace C_p and γ by measured values of the velocity of sound U. This may be achieved by rearranging equations (2.1.11), (12.3.33) and (12.3.34) to give

$$U^2 = (\gamma-1)C_p/T\theta^2 \qquad (12.3.58)$$

A plot of $\log(T\mu'_m/U^2)$ versus $1000/T$ then yields ΔH.

This procedure is inaccurate when $\Delta G < 3RT$, and the further assumption is then required that $\Delta S = 0$, that is $\Delta G = \Delta H$. Under these conditions

$$C_i = R\left(\frac{\Delta H}{RT}\right)^2 \frac{\exp(-\Delta H/RT)}{[1+\exp(-\Delta H/RT)]^2} \qquad (12.3.59)$$

and

$$\mu'_m = \frac{\pi(\gamma-1)R}{2C_p}\left(\frac{\Delta H}{RT}\right)^2 \frac{\exp(-\Delta H/RT)}{[1+\exp(-\Delta H/RT)]^2} \qquad (12.3.60)$$

ΔH can be evaluated from a series of values of μ'_m by an iterative process.

A plot of equation (12.3.59) with $\Delta S = 0$ is shown in Figure 12.3.1, from which it can be seen that C_i has a maximum at $\Delta H/RT = 2.4$. Hence if $\mu'_m C_p/(\gamma-1)$ is found experimentally to increase with increasing temperature, $\Delta H/RT > 3$ and equation (12.3.57) can be used to obtain ΔH and ΔS. If $\mu'_m C_p/(\gamma-1)$ decreases with increasing temperature, then ΔS must be assumed to be zero and ΔH estimated from equation (12.3.60). It sometimes happens that $\mu'_m C_p(\gamma-1)$ passes through a maximum in the experimentally accessible temperature range. The temperature at which this occurs is very close to that at which C_i has a maximum in Figure 12.3.1, and ΔH can be obtained from the relation

$$\Delta H = 2.4RT \qquad (12.3.61)$$

In many cases a molecule has several internal states of the same energy, and the expressions for the relaxing heat capacity must be altered to allow

8

for this. If g_1 and g_2 are the degeneracies of the lower and upper levels, equations (12.3.52) may be generalized to give

$$C_i = \frac{g_2}{g_1} R \left(\frac{\Delta H}{RT}\right)^2 \frac{\exp(-\Delta G/RT)}{[1+(g_2/g_1)\exp(-\Delta G/RT)]^2} \qquad (12.3.62)$$

Corresponding relations exist for the other equations for C_p and μ'_m. The maximum value of C_i occurs at different values of $\Delta H/RT$ for different

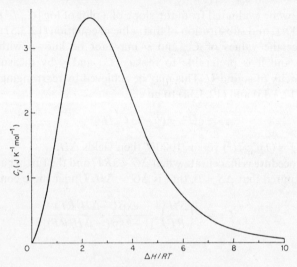

FIGURE 12.3.1. The incremental specific heat C_i for a two-state equilibrium with $\Delta S = 0$.

values of g_1 and g_2. This offers a useful method of determining the degeneracy of the upper and lower states and of identifying these with particular molecular conformations.

12.4   Derivatives of Alkanes

The simplest type of rotational isomerism arises from the alternative orientations of substituents about the $C-C$ single bond in ethane and its derivatives. In ethane itself the three staggered forms are in a lower energy state than the three eclipsed forms, the difference in energy between the two conformations being 12 kJ mol^{-1}. As the CH_3 groups rotate with respect to each other, their hydrogen atoms become alternately staggered and eclipsed, and this gives rise to a barrier to the internal rotation. The exact source of this energy barrier remains uncertain, since electrostatic repulsion or purely steric van der Waals' repulsion of the hydrogen atoms in the eclipsed position is too small to account for the observed barrier height. It is likely that the barrier arises from the repulsion of the C—C and C—H orbitals, and various mechanisms for this have been proposed (Lowe, 1968).

Figure 12.4.1 gives a convenient representation of the conformations of ethane: the C—C bond is assumed to lie in the plane of the paper and the

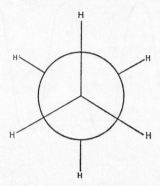

FIGURE 12.4.1. Staggered conformation of ethane in the Newman projection.

substituents at the apices of the two triangles. The potential energy barrier $V(\phi)$ hindering rotation may be represented as a function of angle of rotation ϕ as in Figure 12.4.2. The three staggered conformations of ethane are of equal energy, that is ΔH in equation (12.3.53) is zero. Thus no ultrasonic absorption would be expected from rotational isomerization in ethane.

Substitution of at least one hydrogen atom on each carbon atom leads to a potential energy function which is no longer symmetrical. The larger energy barriers now found contain steric contributions together with electrostatic interactions between polar substituents.

As an example of the determination of barriers to internal rotation in liquids, we consider the results of Lamb (1960) in 1,1,2-trichloroethane. In this liquid there exists a dynamic equilibrium between the three conformations shown in Figure 12.4.3. The passage of an ultrasonic wave perturbs this equilibrium, and ultrasonic absorption is observed since ΔH is not zero. The absorption in this liquid is shown as a function of temperature in Figure 12.4.4. Typical graphs of the absorption per wavelength μ' resulting from the relaxing process are plotted against frequency in Figure 12.4.5 together with the best fit of the equation

$$\mu = \frac{A\omega\tau}{1+\omega^2\tau^2} + B \tag{12.4.1}$$

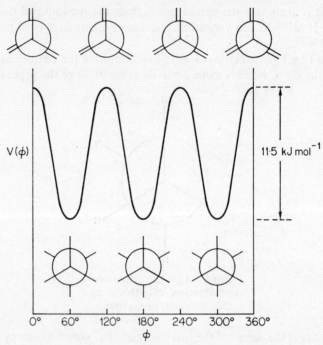

FIGURE 12.4.2. Potential energy $V(\phi)$ as a function of angular displacement ϕ for internal rotation in ethane.

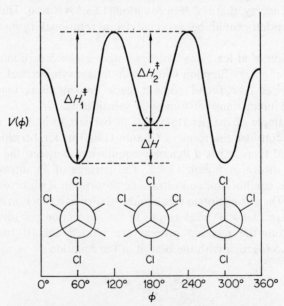

FIGURE 12.4.3. Potential energy $V(\phi)$ as a function of angular displacement ϕ for internal rotation in 1,1,2-trichloroethane.

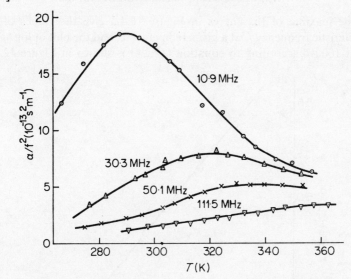

FIGURE 12.4.4. Ultrasonic absorption as a function of temperature in 1,1,2-trichloroethane: ○, 10·9 MHz; △, 30·3 MHz; ×, 50·1 MHz; ▽, 111·5 MHz (after Lamb, 1960).

to the experimental points at each temperature. The first term of equation (12.4.1) represents the relaxational absorption of equation (12.3.36), while B is the absorption resulting from higher frequency relaxations, viscosity,

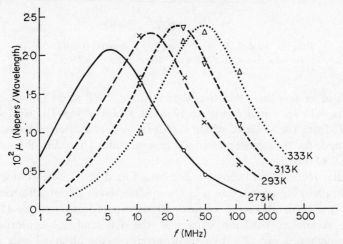

FIGURE 12.4.5. Ultrasonic absorption per wavelength arising from rotational isomeric relaxation in 1,1,2-trichloroethane: curves calculated from equation (12.4.1): —, ○, 273 K; ---, ×, 293 K; ---, ▽, 313 K; ···, △, 333 K (after Lamb, 1960).

etc. The maxima of the curves in Figure 12.4.5 give the values of the characteristic frequency f_c at a given temperature, and the plot of $\log(f_c/T)$ against $1000/T$ according to equation (12.2.4) is shown in Figure 12.4.6.

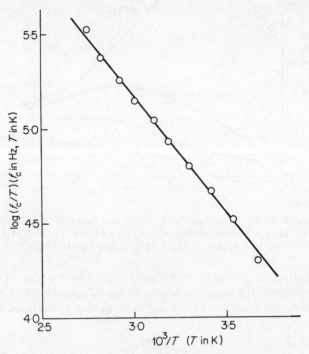

FIGURE 12.4.6. Plot of $\log(f_c/T)$ versus $1000/T$ for 1,1,2-trichloroethane (after Lamb, 1960).

The slope of this line gives the barrier height $\Delta H_b^{\neq}$ as 24 kJ mol^{-1}, and $\Delta S_b^{\neq}$ as -11 JK^{-1} mol^{-1}. Figure 12.4.7 is a plot of $\log(T\mu_m/U^2)$ versus $1000/T$, from the slope of which ΔH may be found after allowance has been made for the degeneracy of the ground state. The value obtained for ΔH is 9 kJ mol^{-1}.

Similar relaxational behaviour has been found in 1,1,2-tribromoethane (Lamb, 1960) but no relaxation has been observed in 1,1,1-trichloroethane where the three staggered conformations are of equal energy (Lamb, 1965). A similar situation occurs in the 1,1- and 1,2-dichloro- and dibromoethanes, excess ultrasonic absorption being observed only in the 1,2-compounds (Piercy, 1965). This provides evidence that the relaxation process observed is indeed the result of restricted internal rotation. Confirmation that an intramolecular process is being studied is found in the work of Piercy (1965) on solutions of 1,2-dichloro- and 1,2-dibromoethane in ether and acetone. He observed that the absorption per wavelength was proportional to the concentration of the solute and that the characteristic

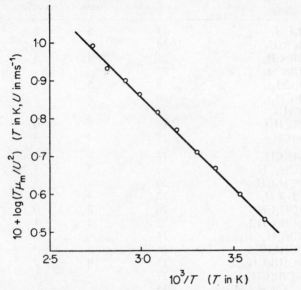

FIGURE 12.4.7. Plot of log $(T\mu_m/U^2)$ versus $1000/T$ for 1,1,2-trichloroethane (after Lamb, 1960).

frequency of the relaxation was approximately constant at different solute concentrations.

Rotational isomerization has been studied in a number of substituted alkanes, and some representative results are given in Table 12.4.1. Only

TABLE 12.4.1. Energy parameters for rotational isomerization of alkane derivatives

Liquid	$\Delta H_b^{\neq}$ (kJ mol^{-1})	ΔH (kJ mol^{-1}) Ultra-sonic	Spectro-scopic	Reference*
CH_2ClCH_2Cl	13	—	—	1
CH_2BrCH_2Br	18	—	—	1
$CHCl_2CH_2Cl$	24	9	4	2, 3, 4, 5
$CHBr_2CH_2Br$	27	7	4	2, 3, 5
$CHBr_2CHBr_2$	18	4	3	6, 7
$CBrF_2CBrF_2$	25	6	4	7, 8
CCl_2FCCl_2H	35	—	—	9

TABLE 12.4.1 (contd.)

Liquid	$\Delta H_b^{\neq}$ (kJ mol^{-1})	ΔH (kJ mol^{-1})		Reference*
		Ultra-sonic	Spectro-scopic	
$CH_2BrCH_2CH_3$	15	5	2	2, 3, 7
$CH_2ClCHClCH_3$	20	5	—	2, 3
$CH_2BrCHBrCH_3$	21	4	3	2, 3, 7
$CH_2ClCHBrCH_3$	23	—	—	10
$CH_2ClCH(CH_3)_2$	16	5	2	11
$CH_2BrCH(CH_3)_2$	20	3	1	11
$CH_2ICH(CH_3)_2$	23	—	—	11
$CH_2ClCCl(CH_3)_2$	19	8	0	11
$CH_2BrCBr(CH_3)_2$	23	4	3	11
$CH_3CH_2CH_2CH_3$	12	—	3	12, 13
$CH_3CHClCH_2CH_3$	19	5	3	11
$CH_3CHBrCH_2CH_3$	20	8	3	11
$CH_3CHICH_2CH_3$	22	4	—	11
$(CH_3)_2CHCH_2CH_3$	20	3	0	14, 7
$(CH_3)_2CHCH(CH_3)_2$	12	4	1	14, 7
$CH_2BrCHBrCH_2CH_3$	14	—	—	10
$CH_3CHClCHClCH_3$	21	5	—	15
$CH_3CHBrCHBrCH_3$	15	8	2	15
$(CH_3)_2CBrCBr(CH_3)_2$	27	—	6	16
$CH_3CH(C_6H_5)CH_2CH_3$	15	—	—	17
$CH_3CH_2CH_2CH_2CH_3$	14	—	2	12, 18
$(CH_3)_2CHCH_2CH_2CH_3$	16	4	—	19
$CH_3CH_2CH(CH_3)CH_2CH_3$	17	4	—	19
$CH_3CHClCH_2CH_2CH_3$	18	—	—	10
$CH_3CHBrCH_2CH_2CH_3$	20	—	—	10
$CH_3CH_2CHClCH_2CH_3$	14	5	—	10
$CH_3CH_2CHBrCH_2CH_3$	15	5	—	10
$CH_2BrCHBrCH_2CH_2CH_3$	15	5	—	10
$CH_3(CH_2)_4CH_3$	9	—	2	12, 18
$CH_3CHBr(CH_2)_3CH_3$	18	6	—	10
$CH_3CH_2CHBr(CH_2)_2CH_3$	15	—	—	10
$CH_3(CH_2)_5CH_3$	12	—	—	12

* References

1. Piercy (1965).
2. Padmanaban (1960).
3. Lamb (1960).
4. Crook and Wyn-Jones (1969).
5. Heatley and Allen (1969).
6. Krebs and Lamb (1958).
7. Sheppard (1959).
8. Petrauskas (1959).
9. Pethrick and Wyn-Jones (1968).
10. Thomas, Wyn-Jones and Orville-Thomas (1969).
11. Wyn-Jones and Orville-Thomas (1968).
12. Piercy and Rao (1967a).
13. Szasz, Sheppard and Rank (1948).
14. Young and Petrauskas (1956).
15. Wyn-Jones (1963).
16. Park and Wyn-Jones (1968).
17. Dexter (1970).
18. Sheppard and Szasz (1949).
19. Chen and Petrauskas (1959).

substances which have been studied by the acoustic technique are included in this table, since other compilations of energy parameters are available (Lowe, 1968). The experimental results for the ultrasonic absorption in all of these compounds are satisfactorily described by a single relaxation despite the fact that most have a number of different energy levels. This implies either that a number of other relaxation processes occur at frequencies without the available range of ultrasonic frequencies or that the observed relaxation time is an average of all the various processes.

The results of Young and Petrauskas (1956) for 2-methyl butane are shown in Figure 12.4.8. A sufficient range of frequency and temperature

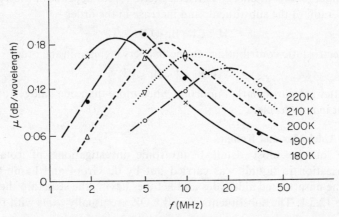

FIGURE 12.4.8. Ultrasonic absorption per wavelength arising from rotational isomeric relaxation in 2-methyl butane: curves calculated from equation (12.4.1): —, ×, 180 K; ---, ●, 190 K; ---, △, 200 K; ···, ▽, 210 K; -·-·-, ○, 220 K (after Young and Petrauskas, 1956).

has been studied to show clearly that the maximum absorption per wavelength μ'_{m} traces out a curve as a function of temperature of similar shape to that of the relaxing heat capacity in Figure 12.3.1. The ground state of 2-methyl butane is doubly degenerate and the maximum value of the relaxing heat capacity C_i occurs at $\Delta H/RT = 2·22$. Since $(\gamma - 1)/C_p$ varies more slowly with temperature than does C_i, it follows that the highest value of μ_{m} is attained when ΔH is close to $2·22RT$. Figure 12.3.1 shows that this occurs at about 185 K, so that $\Delta H = 3·4$ kJ mol^{-1}.

Where comparison is possible, the values of $\Delta H_{\mathrm{b}}^{\neq}$ in Table 12.4.1 are in good agreement with the values obtained by other techniques. In 1-fluoro-1,1,2,2-tetrachloroethane, for example, $\Delta H_{\mathrm{b}}^{\neq}$ is found to be 35 kJ mol^{-1} by the acoustic technique, 37 kJ mol^{-1} by spectroscopy and

31 kJ mol^{-1} by n.m.r. (Pethrick and Wyn-Jones, 1968). There are considerable discrepancies in the values of ΔH in Table 12.4.1, however, the ultrasonic values being substantially larger than the spectroscopic values in many cases. As there are few approximations involved in the spectroscopic determination of energy differences, the ultrasonic values are probably in error. This arises mainly from the assumption that the volume difference ΔV between the rotational isomers is small, and is discussed more fully in Section 12.9.

The values of $\Delta H_b^{\neq}$ in Table 12.4.1 are in good agreement with the arguments of Mizushima (1954) concerning the effect of steric and electrostatic interactions in energy barriers. Steric forces depend on the van der Waals' radii of the substituent, and increase in the order

$$H < Cl < Br < CH_3 < I$$

The electrostatic contributions to the barriers increase in the order

$$CH_3 < I < Br < Cl$$

The latter are particularly important when more than one halogen atom is present in a molecule.

12.5 Aldehydes and Ketones

One of the most detailed ultrasonic investigations of rotational isomerization in liquids was carried out by de Groot and Lamb (1957) on some unsaturated aldehydes and ketones having the structure shown in Figure 12.5.1. The substituents X and COZ are usually *trans* with respect

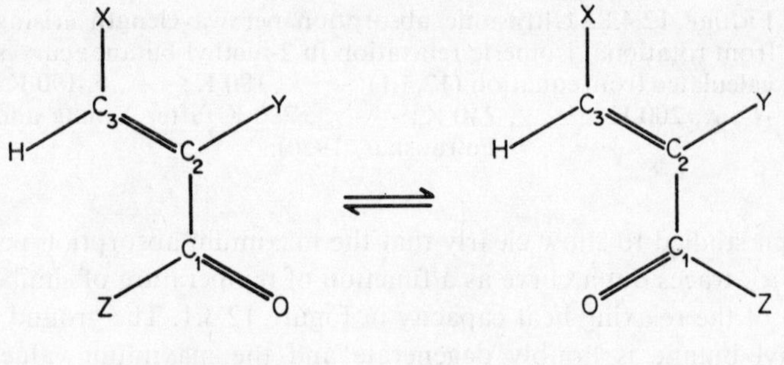

s-trans　　　　　　　　　　*s-cis*

FIGURE 12.5.1. *s-trans* and *s-cis* conformations of unsaturated ketones.

to the $C_2=C_3$ double bond. Conjugation of the $C_1=O$ and $C_2=C_3$ bonds imparts partial double bond character to C_1-C_2 and stabilizes the molecules in the *s-trans* or *s-cis* planar conformations. By varying the substituents X, Y and Z the barrier to rotation about the C_1-C_2 bond is altered and the rate of interconversion of the conformations changed. Such a two-state system should lead to ultrasonic relaxation.

In Figure 12.5.2 the results of de Groot and Lamb (1957) for α/f^2 in crotonaldehyde $CH_3CH:CHCHO$ at three temperatures are shown. The

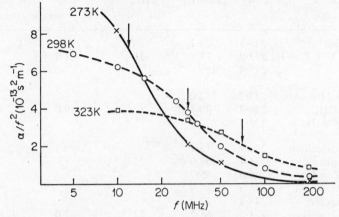

FIGURE 12.5.2. Ultrasonic absorption in crotonaldehyde $CH_3CH:CHCHO$: —, ×, 273 K; ---, ○, 298 K; ---, □, 323 K: curves calculated from equation (2.5.22): arrows indicate f_c (after de Groot and Lamb, 1957).

characteristic frequency of the relaxation increases with temperature, and the value of $\Delta H_b^{\neq}$ is 23 kJ mol^{-1}. From the change in the maximum absorption per wavelength μ'_m with temperature, ΔH is calculated to be 8·1 kJ mol^{-1}. The results for a number of other liquids are summarized in Table 12.5.1.

In order to correlate these figures with the molecular structures involved, it is necessary to establish whether the *s-cis* or *s-trans* conformation is the preferred one. There are two opposing effects to be considered: one arises from the electrostatic attraction of the partially negatively charged oxygen for the partially positively charged C_3; the other is the repulsive interaction between the π-electrons of the double bonds. The former effect would favour the *s-cis* conformation, while the latter would favour the *s-trans*. Dipole moment studies show that in the aldehydes (Z = H) the *trans* position is of lower energy; in ketones (Z = alkyl), however, the *cis* conformation is sometimes favoured because of steric repulsion between Z and the substituents on C_3 (Bentley and coworkers, 1949).

It is interesting to compare the energy values in Table 12.5.1 for acrolein with those for crotonaldehyde and cinnamaldehyde: in the latter substances the hydrogen atom in the X position of acrolein is replaced by a methyl or phenyl group. These groups are electron-donating and should

TABLE 12.5.1. Characteristic frequencies and energy parameters for rotational isomerization of some unsaturated aldehydes and ketones (after de Groot and Lamb, 1957)

Liquid	T (K)	f_c (MHz)	$\Delta H_b^{\neq}$ (kJ mol^{-1})	$\Delta S_b^{\neq}$ (JK^{-1} mol^{-1})	ΔH (kJ mol^{-1})
Acrolein	248·6	29·1			
(X = Y = Z = H)	272·9	77·8	21	−1·8	8·6
	298·4	176			
Crotonaldehyde	273·2	12·0			
(X = CH$_3$;	298·2	30·3	23	−9·0	8·1
Y = Z = H)	323·3	70·0			
Cinnamaldehyde	298·3	15·7			
(X = phenyl;	323·4	36·0	24	−13	6·3
Y = Z = H)	249·0	65·2			
Methacrolein	248·4	22·2			
(X = Z = H;	273·2	65·2	22	+2·0	13
Y = CH$_3$)	298·1	174			
Furacrolein	333·1	13·5			
(X = furyl;	352·6	23·0	21	−36	5·0
Y = Z = H)	373·0	31·4			
2-Ethyl-3-propyl acrolein	298·1	24·6			
(X = propyl;			—	—	—
Y = ethyl; Z = H)	323·3	56·0			
Hexyl cinnamic aldehyde	298·2	18·2			
(X = phenyl;			—	—	—
Y = hexyl; Z = H)	323·4	35·8			
Methyl vinyl ketone	248·4	80·6			
(X = Y = H;			—	—	≪6
Z = CH$_3$)	273·1	178			
Methyl isopropenyl ketone	248·4	26·4			
(X = H;			—	—	≪6
Y = Z = CH$_3$)	273·1	59·6			
Furfural acetone	323·3	9·8			
(X = furyl;			—	—	≪6
Y = H; Z = CH$_3$)	347·9	23·2			

strengthen the conjugation of the C_1—C_2 bond and hence increase the rotational barrier $\Delta H_b^{\neq}$. Table 12.5.1 shows that the barrier increases as the electron-donating ability of the substituent increases, while the characteristic frequency for traversing the barrier decreases. The increased charge separation in these molecules resulting from the inductive effect leads to stronger intramolecular electrostatic forces which stabilize the higher energy *s-cis* conformation.

Substitution of a similar electron-donating group in the Y position should have little effect on the conjugation or on the energy barrier. Hence the characteristic frequencies of acrolein and methacrolein and of cinnamaldehyde and hexyl cinnamic aldehyde are very similar at a given temperature. The *s-trans* conformation in methacrolein is stabilized when compared with acrolein. This is thought to arise from an attraction between the carbonyl oxygen and the methyl group.

In the three ketones in Table 12.5.1, the aldehydic hydrogen is replaced by a methyl group. Although the characteristic frequency of methyl vinyl ketone is larger than that of acrolein, f_c of methyl isopropenyl ketone and of furfural acetone is similar to that of the corresponding aldehyde. This suggests that the barrier to rotation about the C_1—C_2 bond is little affected by replacing the C_1 hydrogen atom by a methyl group. de Groot and Lamb (1957) found that μ_m decreases with increasing temperature indicating a value of $\Delta H \ll 6$ kJ mol^{-1}. This small enthalpy difference is probably the result of steric repulsion between the C_1 methyl group and the C_3 hydrogen atom.

An important exception to these conclusions is mesityl oxide (Figure 12.5.3) in which no ultrasonic relaxation was observed over the entire

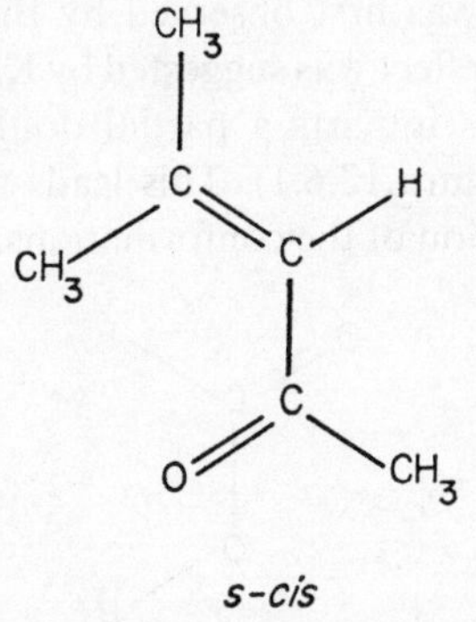

FIGURE 12.5.3. Mesityl oxide.

s-cis

available range of frequency and temperature. Molecular models show that the *s-trans* conformation of this substance is inaccessible as it is completely blocked sterically. This provides useful confirmation that the origin of the observed relaxational behaviour in such substances lies in the equilibrium between *s-trans* and *s-cis* isomers.

de Groot and Lamb (1957) also investigated briefly some other carbonyl compounds. In acetaldehyde where the alternative conformations are equal in energy, no relaxation was observed. The same was true of benzaldehyde, anisaldehyde and acetophenone. In propionaldehyde and *n*-butyraldehyde, however, the various conformations differ in energy and ultrasonic relaxation was observed: this occurred at high frequencies and was accessible at low temperatures only, indicating that the absence of conjugation leads to a greatly decreased rotational barrier. This is borne out by n.m.r. and microwave studies of propionaldehyde which show that $\Delta H_b^{\neq}$ is 5 kJ mol^{-1}, the *cis* conformation being more stable by 4 kJ mol^{-1} (Abraham and Pople, 1960; Butcher and Wilson, 1964).

Ultrasonic relaxation was also observed by de Groot and Lamb (1957) in furan- and thiophene-2-aldehyde, the lower characteristic frequency of the former indicating stronger conjugation between the aldehydic group and the ring. Replacement of the aldehydic hydrogen by a CH_3 group leads to steric hindrance, a decrease in conjugation and an increase in the characteristic frequency of the relaxation process. Pethrick and Wyn-Jones (1969) have made a careful comparison of the barriers in the liquid phase determined by ultrasonics and n.m.r. in these aldehydes. Within experimental error there is excellent agreement, the barrier heights $\Delta H_b^{\neq}$ being 40 kJ mol^{-1} in furan-2-aldehyde and 43 kJ mol^{-1} in thiophene-2-aldehyde. These barriers are much greater than those of the unsaturated aldehydes because of the increased conjugation of the heterocyclic ring.

12.6 Esters and Vinyl Ethers

Ultrasonic relaxation in liquid esters was first observed by Biquard in 1936, and the accepted explanation of the effect was suggested by Karpovich (1954). Delocalization of the π-electrons imparts a partial double bond character to the central C—O bond (Figure 12.6.1). This leads to a substantial energy barrier to the interconversion of the conformations in which

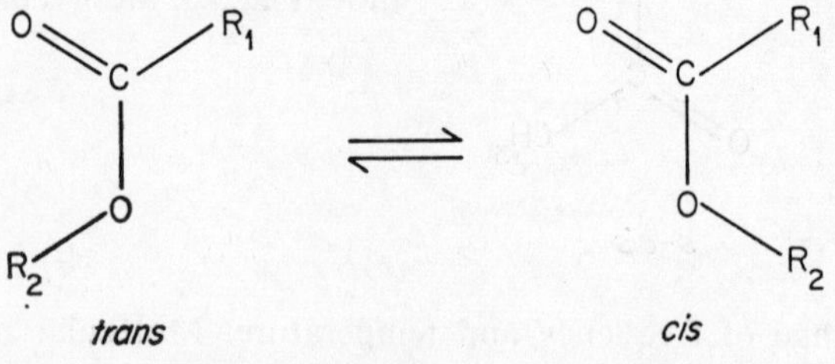

FIGURE 12.6.1. The two conformations of esters, with alkyl groups *trans* and *cis*.

the alkyl groups are *trans* and *cis*. Most physical measurements are unable to detect the higher energy *cis* conformation.

Figure 12.6.2 gives the results of Subrahmanyam and Piercy (1965a) for the ultrasonic absorption per wavelength in pure ethyl formate along with the theoretical curves for a single relaxation process. The value of $\Delta H_b^{\neq}$ is 33 ± 2 kJ mol^{-1}. Table 12.6.1 contains some results for the energy parameters in esters determined by the acoustic method.

The chief difficulty in studying ultrasonic relaxation in esters is that the characteristic frequencies lie in the range 100 kHz–10 MHz where the experimental accuracy is low. This is the main cause of the discrepancies between the results of different workers in Table 12.6.1. Piercy and Subrahmanyam (1965a) overcame this difficulty by working well above the normal liquid boiling point and maintaining the liquid state by the

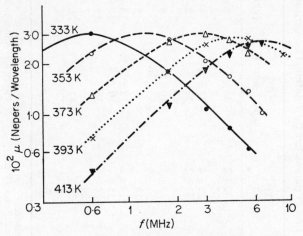

FIGURE 12.6.2. Ultrasonic absorption per wavelength arising from rotational isomeric relaxation in ethyl formate: curves calculated from equation (12.4.1): —, ●, 333 K; ---, ○, 353 K; ---, △, 373 K; ···, ×, 393 K; -·--, ▼, 413 K (after Subrahmanyam and Piercy, 1965a).

application of pressure. The characteristic frequencies were then brought into the accurate experimental range (> 5 MHz). This procedure is justified since Slie and Litovitz (1963) have found that the application of pressure does not alter the relaxation frequency of ethyl acetate at a given temperature (Table 12.6.2).

Several other studies of relaxation in esters have been reported. The work of Karpovich (1954) was conducted at a single temperature and no

TABLE 12.6.1. Energy parameters for rotational isomerization of esters

Liquid	State	$\Delta H_b^{\neq}$ (kJ mol^{-1})	$\Delta S_b^{\neq}$ (JK^{-1} mol^{-1})	ΔH (kJ mol^{-1})	ΔS (JK^{-1} mol^{-1})	Reference*
Methyl formate	Pure	32 ± 2	-36 ± 7	$8 - 12$	—	1
	Pure	32	-8	2	4	2
	6% in xylene	29 ± 4	-25 ± 12	$8 - 12$	—	1
Ethyl formate	Pure	33 ± 2	-37 ± 5	$8 - 12$	—	1
	Pure	26	-40	2	8	2
	6% in xylene	33 ± 3	-40 ± 8	$8 - 12$	—	1
	6% in nitrobenzene	31 ± 4	-33 ± 10	—	—	1
n-Propyl formate	Pure	28	-28	15	7	2
i-Propyl formate	Pure	24 ± 2	-18 ± 5	15 ± 2	6 ± 3	3
Methyl acetate	Pure	25	-10	18	0	2
Ethyl acetate	Pure	18	-36	19	-4	2
	Pure	24	—	7	—	4
Methyl propionate	Pure	20	-42	21	7	2
Ethyl propionate	Pure	5	-77	24	6	2
Ethyl n-butyrate	Pure	2	—	—	—	2

* References
1. Subrahmanyam and Piercy (1965a).
2. Bailey and North (1968).
3. Piercy and Subrahmanyam (1965a).
4. Slie and Litovitz (1963).

energy parameters are available: relaxation was observed in methyl and ethyl formates and acetates as well as in ethyl acetoacetate, methyl benzoate and acetic anhydride. Tabuchi (1958) studied ethyl formate by measuring dispersion of sound rather than absorption: since the

TABLE 12.6.2. Effect of pressure on characteristic relaxation frequency of liquid ethyl acetate (after Slie and Litovitz, 1963)

T (K)					
313	P (atm)	1	381	760	1000
	f_c (MHz)	22·7	23·1	22·7	23·1
353	P (atm)	1	351	700	1000
	f_c (MHz)	71·2	71·0	69·4	76·2

velocity dispersion is small, his values of $\Delta H_b^{\neq} = 14 \text{ kJ mol}^{-1}$ and $\Delta H = 10 \text{ kJ mol}^{-1}$ are less reliable than those in Table 12.6.1. Pancholy and Mathur (1963) have observed relaxation processes in twenty-four esters, the characteristic frequencies of which lie in the range 2–5 MHz at room temperature.

The most extensive study of esters is that of Bailey and North (1968), and their results are included in Table 12.6.1. The lower energy conformations of a number of different esters lie about 37 kJ mol^{-1} below the transition state. The energy of the higher energy conformation in which the alkyl groups are *cis* to each other rises with increasing number of carbon atoms on both alkyl groups. This observation permits a differentiation among the various proposals which have been made to account for the relative stability of the two conformations. If the *trans* state (Figure 12.6.1) were stabilized by hydrogen bonding between R_2 and the carbonyl oxygen, then the relative stabilities would depend primarily on R_2 and not on R_1. This is not observed. There is also no correlation between the energy parameters and the molecular dipole moment so that dipole stabilization of the *trans* state which has the lower dipole moment is not important. The results are in complete agreement with the theory that the energy differences arise from steric repulsion of the alkyl groups in the higher energy *cis* conformations. This is supported by the work of Burundukov and Yakovlev (1968, 1969) on a series of *n*-alkyl formates.

A second relaxation region has been detected by Piercy and Subrahmanyam (1965a) in *i*-propyl formate at frequencies well above the main relaxation region. This was ascribed to rotation about the other C—O bond as in Figure 12.6.3. The relaxation frequency was too high for

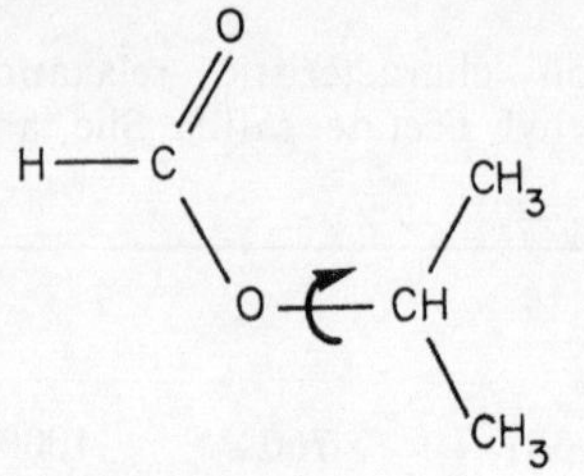

FIGURE 12.6.3. Internal rotation in *i*-propyl formate which gives rise to a high frequency ultrasonic relaxation process.

the values of the energy parameters to be evaluated. Such energy barriers have been studied by n.m.r. and microwave spectroscopy, however, a typical barrier being 5 kJ mol^{-1} for rotation of the CH_3 in methyl formate (Curl, 1959).

Nozdrev (1965) has also detected two relaxation regions in several esters. In this case the curves of μ' versus frequency are too narrow to be described by equation (12.4.1), and so the validity of the two relaxation regions must be considered doubtful.

A relaxation process similar to the main relaxation in esters was observed in some vinyl ethers by de Groot and Lamb (1957). As in the esters, the observed energy barrier must arise from the partial double bond character of the central C—O bond (Figure 12.6.4). Both *cis* and *trans* forms of such

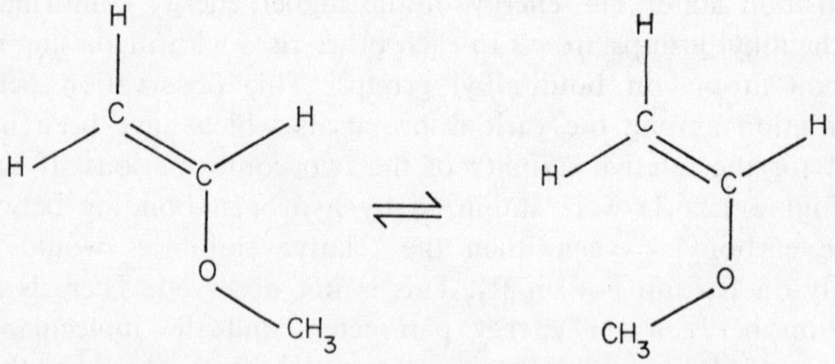

FIGURE 12.6.4. Internal rotation in methyl vinyl ether.

ethers have been observed in the liquid by infrared spectroscopy (Owen and Sheppard, 1963). No energy parameters have been obtained for these ethers by the ultrasonic method.

12.7   Cyclic Compounds

Cyclohexane can exist in two conformations, a boat form and a chair form (Figure 12.7.1). Neither has angle strain, but in the boat form there

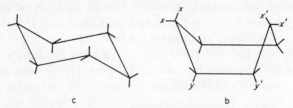

FIGURE 12.7.1. Chair (c) and boat (b) conformations of cyclohexane.

is bond opposition strain (y–y' in Figure 12.7.1) and steric repulsion of the pair of hydrogen atoms at the top of the boat (x–x' in Figure 12.7.1). In the chair form there is less steric repulsion, and it is considerably more stable than the boat. The enthalpy difference between the two forms is about 23 kJ mol^{-1} while the energy barrier $\Delta H_b^{\neq}$ is about 17 kJ mol^{-1} (Eliel, 1962). As a result, only about one molecule in a thousand is in the boat form at 298 K and the conversion from boat to chair is extremely rapid. In agreement with this, no ultrasonic relaxation has been observed in liquid cyclohexane at frequencies up to 100 MHz: at 298 K there is some evidence for the onset of relaxation at 200 MHz but the matter has not been investigated thoroughly (Karpovich, 1954; Heasell and Lamb, 1956b).

In the chair conformation of cyclohexane there are two types of bonds, namely those which are parallel to the principal axis of the ring and are called 'axial' and those which make an angle of 109° with this axis and are termed 'equatorial'. Thus there exist two conformational isomers of the chair form of a monosubstituted cyclohexane (Figure 12.7.2), and these

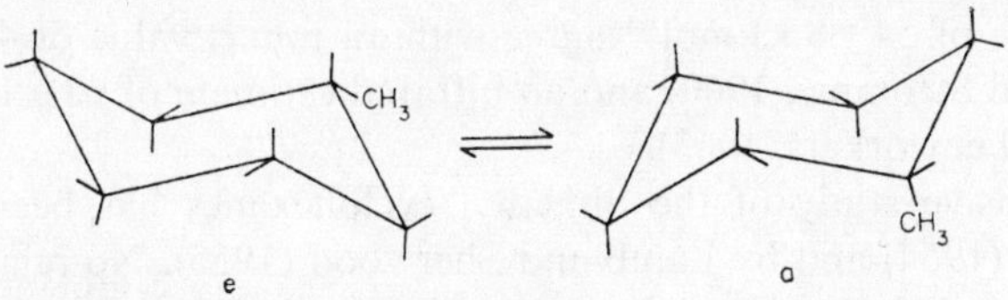

FIGURE 12.7.2. Conformations of methyl cyclohexane:
e, equatorial; a, axial.

can be interconverted by a process of ring inversion in which no bonds are broken. There is greater steric interaction in the axial than in the equatorial form, and the resulting energy difference should lead to ultrasonic relaxation.

The first study of liquid cyclohexane derivatives was that of Karpovich (1954) using the reverberation technique at room temperature. Relaxation was observed with a characteristic frequency of about 100 kHz in methyl cyclohexane, cyclohexanol, cyclohexylamine and dicyclohexylamine. Similar behaviour was found by Lamb and Sherwood (1955) in methyl and ethyl cyclohexane, and methyl cyclohexane was further studied by Hall (1959) and Pedinoff (1962). The most extensive results for mono-substituted cyclohexanes are those of Piercy (1961) who covered the frequency range 0·6–25 MHz using the streaming and pulse techniques.

TABLE 12.7.1. Energy parameters for chair–chair inversion in cyclic compounds

Liquid	State	$\Delta H_b^{\neq}$ (kJ mol^{-1})	ΔH (kJ mol^{-1})	Reference[*]
Methyl cyclohexane	Pure	43	12	1
	In xylene	46	15	2
	In nitrobenzene	45	15	2
	In n-butanol	45	—	2
Chlorocyclohexane	In xylene	50	—	2
Bromocyclohexane	In xylene	50	—	2
Trimethylene sulphite	Pure	23	5	3
4-Methyl trimethylene sulphite	Pure	20	5	3
4-Methyl-1,3-dioxan	Pure	39	11	4
4-Phenyl-1,3-dioxan	Pure	19	22	4

[*] References
1. Piercy and Subrahmanyam (1965b).
2. Piercy (1961).
3. Eccleston and coworkers (1970).
4. Hamblin, White and Wyn-Jones (1969).

Table 12.7.1 contains the values of the energy parameters of three mono-substituted cyclohexanes. The values of $\Delta H_b^{\neq}$ for chloro- and bromo-cyclohexane of 54 ± 8 kJ mol^{-1} agree with an n.m.r. value of 45 kJ mol^{-1} (Reeves and Strømme, 1960) and an infrared estimate of 63 ± 19 kJ mol^{-1} (Pentin and coworkers, 1963).

An extensive study of the dimethyl cyclohexanes has been made by Karpovich (1954) and by Lamb and Sherwood (1955). No relaxation was observed in cis-1,2-, trans-1,3- or cis-1,4-dimethyl cyclohexane. Each of these substances has one axial and one equatorial methyl group, and on ring inversion this also yields one axial and one equatorial substituent. As there is no energy difference between the two conformations, no ultrasonic relaxation would be expected. In the case of trans-1,2-, cis-1,3- and trans-1,4-dimethyl cyclohexane, however, ring inversion converts two equatorial methyl groups into two axial groups, and the resulting energy difference between the two conformations should lead to ultrasonic relaxation. This has been observed in the *trans*-1,2- and *trans*-1,4- compounds, the characteristic frequencies being 120 and 150 kHz respectively at 289 K. No relaxation was observed in *cis*-1,3-dimethyl cyclohexane in the frequency range studied (150–650 kHz). In this case the axial methyl groups experience considerable steric interaction and the barrier $\Delta H_b^{\neq}$ is decreased. Hence ultrasonic relaxation would be expected to occur at higher frequencies in *cis*-1,3-dimethyl cyclohexane. The same behaviour should be seen in 1,1,3-trimethyl cyclohexane.

Karpovich (1954) also observed relaxation in *o*-methyl cyclohexanone arising from an axial–equatorial equilibrium, but not in cyclohexanone itself. In cyclohexene a relaxation process was found which was attributed to an equilibrium between the pseudo-chair and pseudo-boat conformations. Since the pseudo-boat of cyclohexene has less steric interaction than has the boat in cyclohexane, a lower characteristic frequency would be expected in the former. Experimentally Karpovich (1954) found that f_c of cyclohexene was 80 kHz at 309 K. A similar result was obtained by Pedinoff (1959). These conclusions are contradicted by the acoustic streaming results of Subrahmanyam and Piercy (1965b) which show that α/f^2 in cyclohexene is independent of frequency in the range 600 kHz–200 MHz.

The equilibrium between the axial and equatorial chair conformations of substituted 1,3-dioxans has been studied by Hamblin, White and Wyn-Jones (1968, 1969). In the 4-methyl compound the characteristic frequency is 7·8 MHz at room temperature, while that of 4-phenyl 1,3-dioxan is 10·7 MHz: these frequencies are unchanged in *p*-xylene solution. Ultrasonic relaxation has also been observed in trimethylene sulphite and several methyl-substituted derivatives by Eccleston and coworkers (1970). This was attributed to a perturbation of the equilibrium between two chair conformations in which the exocyclic S=O group is either axial or equatorial. $\Delta H_b^{\neq}$ is typically 20 kJ mol^{-1} in the cyclic sulphites, a value which is substantially less than in the cyclohexanes in Table 12.7.1.

12.8   Tertiary Amines

Measurements of ultrasonic absorption in triethylamine (Heasell and Lamb, 1956a) and tri-*n*-butylamine (Krebs and Lamb, 1958) have revealed a relaxation process which has been attributed to rotational isomerization. Figure 12.8.1 shows the three possible conformations of triethylamine. Krebs and Lamb (1958) concluded from a study of molecular models that conformation c of Figure 12.8.1 is precluded because of the considerable steric interaction. The ethyl groups have more freedom of movement in conformation b than in a. Since the experimental value of ΔS is positive

(Table 12.8.1), conformation a must be the state of lowest energy in triethylamine, and the attractive force of the nitrogen atom for the ethyl groups must outweigh the steric repulsions.

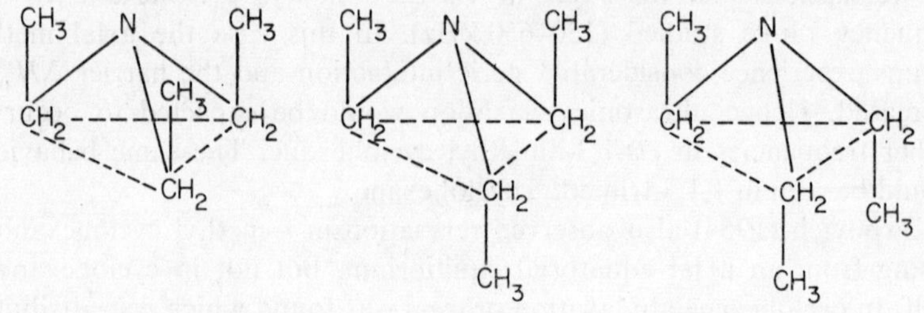

FIGURE 12.8.1. Conformations of triethylamine.

TABLE 12.8.1. Energy parameters for rotational isomerization of amines

Liquid	$\Delta H_b^{\neq}$ (kJ mol^{-1})	ΔH (kJ mol^{-1})	ΔS (JK^{-1} mol^{-1})	Reference*
Triethylamine	28	14	20	1
Tri-*n*-propylamine	19	5	5	2
Tri-*n*-butylamine	18	4	−4	3
Tri-*n*-pentylamine	20	5	−4	2
Tri-isopentylamine	13	5	−5	2
Tri-*n*-hexylamine	18	5	−5	2

* References
1. Heasell and Lamb (1956a).
2. Williams and coworkers (1968).
3. Krebs and Lamb (1958).

Litovitz and Carnevale (1958) have examined the pressure dependence of the relaxation frequency of triethylamine and their results are given in Table 12.8.2. The characteristic frequency is independent of the applied pressure. This supports the intramolecular interpretation of the relaxation

TABLE 12.8.2. Effect of pressure on the characteristic relaxation frequency of liquid triethylamine at 273 K (after Litovitz and Carnevale, 1958)

P (atm)	1	2000	3280
f_c (MHz)	31·2	32·8	31·2

process, and shows that the volume change in reaction ΔV is close to zero in this substance.

Table 12.8.1 contains some representative values of the energy parameters in tertiary amines. ΔS changes from a positive to a negative value as the length of the alkyl chain increases: this suggests that isomer b becomes more stable at large chain lengths. The energy barriers $\Delta H_b^{\neq}$ are reasonably constant at about 19 kJ mol^{-1} in those amines with side-groups larger than ethyl. This is consistent with rotation about a C—N bond.

12.9 Limitations of the Acoustic Method of Studying Rotational Isomerization

A number of assumptions have been made in Sections 12.2 and 12.3 in deriving the relations between the observed ultrasonic absorption and the molecular energy parameters. In determining the energy barrier $\Delta H_b^{\neq}$ the assumptions are:

(a) the process is intramolecular, and may be represented as a two-state system;

(b) the reaction is unimolecular;

(c) $\Delta G_b^{\neq} \ll \Delta G_f^{\neq}$;

(d) $\Delta H_b^{\neq}$ and $\Delta S_b^{\neq}$ are independent of temperature.

In determining the energy difference between isomers ΔH, assumptions (a) and (b) are involved, together with:

(e) the dispersion of velocity in the relaxation region is negligible;

(f) the volume change on reaction ΔV is small;

(g) ΔS and ΔH are independent of temperature;

(h) equilibrium parameters such as heat capacity and expansion coefficient are independent of temperature;

and either

(i) $\Delta G > 3RT$;

or

(j) ΔS is negligible.

We now discuss the importance of these assumptions.

(a) Evidence that the relaxation process is intramolecular has been obtained by working in different solvents or over a range of pressure. Such studies have been made on a number of substances which undergo rotational isomerization, and the characteristic frequency of the relaxation is almost independent of pressure or of the molecular environment. In contrast to this, the characteristic frequencies of intermolecular relaxation processes such as vibrational relaxation (Chapter 11) or structural relaxation (Chapter 10) vary markedly with pressure. Within the limits of experimental error, ultrasonic absorption caused by rotational isomerization may be described by a single relaxation time corresponding to a two-state process.

(b) That the reaction is unimolecular may be verified by studying the rotational isomerization of a substance at different concentrations in a solvent. Little dependence of the characteristic frequency on concentration has been detected (Piercy, 1961, 1965).

(c) This assumption is justified in many cases since only one molecular conformation has been detected by non-ultrasonic methods. Also, the plot of $\log(f_c/T)$ versus $1000/T$ is linear within experimental error. This assumption is not always necessary, however, since in many cases a value of ΔG can be obtained from the ultrasonic results.

(d) and (g) These assumptions are implicit in all methods of determining energy parameters. Since the process being studied is intramolecular, it is unlikely that this assumption is seriously in error.

(e) This assumption could be eliminated by measuring the ultrasonic velocity over a sufficiently wide range of frequency in the relaxation region and modifying the equations of Section 12.3 to include velocity dispersion. This is found to be unnecessary since velocity dispersion does not exceed 1%.

(f) The assumption that $\Delta V C_p/\Delta H V \theta \ll 1$ is one of the major assumptions involved in the determination of energy differences between rotational isomers. One important test of its validity is to measure the characteristic frequency of the relaxing process over a range of pressures at a given temperature. In the case of ethyl acetate (Table 12.6.2) and triethylamine (Table 12.8.2), f_c is independent of pressure, suggesting that the assumption of negligible volume change on reaction is valid. Nevertheless it has been shown that $\Delta V C_p/\Delta H V \theta$ in equation (12.3.53) is not always much smaller than unity, and this assumption can lead to considerable inaccuracies (Wyn-Jones and Orville-Thomas, 1966; Singh, Darbari and Verma, 1967; Crook and Wyn-Jones, 1969).

(h) This assumption is unjustified over large ranges of temperature, although it is likely that over moderate temperature ranges the resulting discrepancy is not serious. A correction may be made if the temperature dependence of the heat capacity and expansion coefficient are known.

(i) and (j) If $\Delta G > 3RT$, the error in ΔG does not exceed 5%. There is usually little information on ΔS, although a reasonable range appears to be $-8 < \Delta S < 8 \text{ JK}^{-1} \text{ mol}^{-1}$. Hence only very approximate values of ΔH can be obtained when ΔH is small.

There is also the important assumption in the derivation of energy differences that the mixing of the two isomers is ideal. That this is incorrect is shown by the different values of ΔH found in different solvents: this arises from the progressive stabilization of the isomer of larger dipole moment as the dielectric constant of the solvent increases (Heatley and Allen, 1969). It is also assumed that the absorption arising from rotational isomerization is additive with the ultrasonic absorption arising from other causes.

We conclude that the assumptions involved in the determination of $\Delta H_b^{\neq}$ are not serious, and hence the barrier heights obtained should agree with those determined in other ways. Unfortunately few data are available for comparison with the ultrasonic values, although the spectroscopic barrier heights in 1,1,2,2-tetrachloroethane (Section 12.4) and chloro- and bromo-cyclohexane (Section 12.7) agree well with the acoustic values.

The ultrasonic determination of energy differences yields much less satisfactory results. As can be seen in Table 12.4.1, the ultrasonic values often differ considerably from those determined by other methods, especially when ΔG is small. Moreover the results from several other techniques are consistent and do not involve such severe assumptions as in the ultrasonic case. Thus ultrasonic values of energy differences obtained from measurements in the pure liquid are suspect.

As Davies and Lamb (1959) have shown, it is inevitable that some assumptions be made in the ultrasonic derivation of energy differences, but there is little experimental evidence to show which assumptions are valid. In conventional ultrasonic work on pure liquids, errors arise mainly from the assumption that ideal mixing of the two isomers occurs and that the volume difference between isomers is zero. An approximation to ideal mixing can be obtained by working in reasonably dilute solutions of the relaxing substance in an inert solvent. Piercy (1961) has found that if the relaxing substance is non-polar, ΔH is little affected by the nature of the solvent. Similarly if a non-polar solvent can be found for a polar substance, ΔH should again be unaffected by the solvent. Even in dilute solution, however, there remains the problem of the volume difference between isomers. By varying the solvent, it is possible to vary $C_p/V\theta$ in equation (12.3.53): if the resulting values of ΔH are independent of solvent, it is evident that ΔV is negligible. Alternatively, it is sometimes possible to make a reasonable estimate of ΔV and so correct for this effect. The best method of checking that $\Delta V = 0$ is by the application of pressure, but this method has not been widely used.

Thus the most promising method of determining energy differences between isomers by the ultrasonic technique is to study the relaxing substance in dilute solution in an inert solvent. As yet, this approach has not been thoroughly tested for substances for which ΔH has been established by other techniques. The main advantage of the ultrasonic method is that it can readily be applied to the determination of energy parameters of substances which cannot be studied by other techniques. For example, in the determination of ΔH most other methods require an equilibrium density of molecules in the higher energy state of at least 10%, corresponding to $\Delta H/RT = 2{\cdot}3$ or $\Delta H = 6$ kJ mol^{-1} at 300 K. The ultrasonic method can provide information when only $0{\cdot}5\%$ of the molecules are in the upper state, corresponding to $\Delta H/RT = 5{\cdot}3$ or $\Delta H = 13$ kJ mol^{-1} at 300 K.

CHAPTER 13

PROPAGATION OF ULTRASONIC
LONGITUDINAL WAVES IN POLYMERS

13.1 Viscoelastic and Structural Relaxation

Meaningful determinations of the propagation constants of ultrasonic longitudinal waves in liquids can only be made at frequencies above 1 MHz (Chapter 3). The molecular theories of the viscoelasticity of polymers outlined in Chapters 7 and 8 show that the period of such an ultrasonic wave is much shorter than many of the relaxation times for the motion of polymer chains. Hence with longitudinal waves of frequency greater than 1 MHz it is only possible to study rapid, localized motions of polymer chains.

An example of this is given by the results of Mikhailov and Tarutina (1960) for the propagation of an ultrasonic longitudinal wave of frequency 10·4 MHz in a 1·5% aqueous solution of gelatin having a viscosity of 10^2 N s m^{-2}. The classical absorption of this solution calculated from equation (2.2.6) would be 8×10^4 m^{-1}, while that of the solvent is 3 m^{-1}. The observed value of the absorption coefficient is 4·8 m^{-1}. Many other studies have confirmed that the ultrasonic absorption of a polymer solution is approximately equal to that of the pure solvent (Mikhailov, 1963, 1964).

This low absorption arises from the small values of the dynamic viscosities of polymer liquids and solutions in the megahertz frequency range. An interesting examination of the relation between viscoelastic and longitudinal relaxation was undertaken by Hunter and Derdul (1967) in some of the polydimethylsiloxane liquids whose viscoelastic behaviour was defined by Barlow, Harrison and Lamb (1964) (Section 8.3). When the measured values of dynamic shear viscosity are substituted into equation (10.3.16), the ultrasonic absorption may be calculated if the dynamic volume viscosity is assumed to be negligible. Figure 13.1.1 gives a comparison of the ultrasonic absorption calculated in this way with the results of Hunter and Derdul (1967) for a polydimethylsiloxane liquid of nominal, room temperature viscosity of $3·5 \times 10^{-4}$ m^2 s^{-1}. The good agreement shows that ultrasonic absorption in the polydimethylsiloxane liquids is determined largely by the dynamic shear viscosity. The contribution of the dynamic volume viscosity appears to be small: this is probably the result of a small volume viscosity in the polydimethylsiloxanes, although it is possible that relaxation of a larger volume viscosity has occurred at frequencies below 1 MHz. No velocity dispersion has been observed in the polydimethylsiloxane liquids in the megahertz frequency range. This is consistent with the existence of lower frequency relaxation processes.

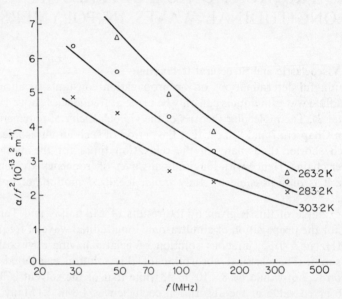

FIGURE 13.1.1. Experimental and predicted values of the ultrasonic absorption for a polydimethylsiloxane liquid of nominal kinematic viscosity 3.5×10^{-4} m^2 s^{-1}. The points are experimental values, and the curves are obtained from equation (10.3.16) using the results of Barlow, Harrison and Lamb (1964) for η'_s and assuming $\eta'_v = 0$. $\triangle$, 263.2 K; $\bigcirc$, 283.2 K; $\times$, 303.2 K (after Hunter and Derdul, 1967).

Alternatively it implies that the relaxational bulk modulus K_2 and rigidity modulus G_∞ are small, since from equations (10.3.13) and (10.3.14)

$$\frac{U_\infty^2}{U_0^2} = \frac{K_0 + K_2 + \frac{4}{3}G_\infty}{K_0} \approx 1 \qquad (13.1.1)$$

The high flexibility of the polydimethylsiloxane chain permits the rapid rearrangement of small chain segments, and this may account for the low dynamic volume viscosity and elastic constants. In polyisobutyl-methacrylate the chain backbone is more rigid: the dynamic volume and shear viscosities are now comparable in size, while K_2 and G_∞ are no longer negligible compared to K_0 (Morita, Kono and Yoshizaki, 1968).

Ultrasonic absorption has also been used to study the motions of polymer segments (cf. Section 8.5). An example of this is the work of Eby (1964) on polyethylene. Hitherto the results of ultrasonic studies of polymers have been similar to those from shear wave studies but have covered a more limited time scale. The main applications of the ultrasonic technique will be in studying local motions of polymer chains to provide information complementary to that obtained from dielectric relaxation and nuclear magnetic relaxation.

13.2 Rotational Isomerization in Polymers

The simplest internal motion of a polymer chain is the rotation of a side-group about the chain backbone. Where the resulting conformations have different energies, the ultrasonic technique can be used to determine the rate of interconversion of the conformations and, with certain assumptions, the energy difference between the conformations.

The study of rotational isomerization in simple liquids using ultrasonic longitudinal waves has been discussed in Chapter 12. Figure 13.2.1 shows the absorption per wavelength μ as a function of frequency for a solution of polystyrene in carbon tetrachloride (Hässler and Bauer, 1969). Although the difference in absorption between the solution and solvent is subject to considerable error, the excess absorption is adequately represented by a single relaxation curve.

This excess absorption is proportional to the concentration of the polymer but independent of molecular weight in polystyrene solutions. Hässler and Bauer (1969) attribute it to an internal rotation of the polystyrene molecule in which rotation of a benzene ring around the chain backbone is accompanied by a rotation of a backbone carbon–carbon bond in the opposite direction ('kink' conformation). From the temperature dependence of the characteristic frequency of the relaxation process, the barrier to the conformational change $\Delta H_b^{\neq}$ was found to be 27 kJ mol^{-1}, while the energy difference between the conformations was 4 kJ mol^{-1}.

Similar results have been found for solutions of polymethylmethacrylate by Nomura, Kato and Miyahara (1969). One difficulty which may complicate the interpretation of such results is that the high frequency 'tail' of the viscoelastic relaxation of the polymer may also contribute to the observed ultrasonic absorption in polymer solutions.

13.3 Helix-coil Transition in Polypeptides

Polypeptides may exist in solution as rigid rods in which a helical molecular conformation is stabilized by hydrogen bonding. If the temperature, solvent or pH of a polypeptide solution are altered, the α-helix may

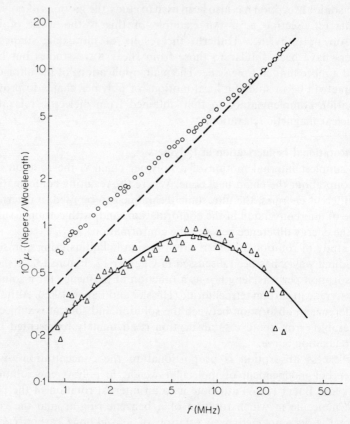

FIGURE 13.2.1. Ultrasonic absorption per wavelength μ in a 5% solution of polystyrene of molecular weight $2\cdot0 \times 10^5$ in carbon tetrachloride at 293·2 K: $\circ$, μ of solution; ---, classical absorption of solvent (equation 2.2.6); $\triangle$, difference between μ of solution and solvent; —, predictions of single relaxation theory (equation 12.4.1) with $f_c = 6\cdot6$ MHz (after Hässler and Bauer, 1969).

break up to form a random coil. Temperature jump techniques suggest that the relaxation time for the transition from helix to random coil is less than 10^{-6} s, and accordingly several investigations of ultrasonic absorption in polypeptide solutions have been carried out.

Figure 13.3.1 shows the ultrasonic absorption in 0·16 M poly-L-lysine in 0·6 M NaCl at 309 K and at various values of pH between 5·1 ($<3\%$ polymer in helical form) to 9·6 (15% in helical form). The solid lines represent the single relaxation equation for a two-state chemical equilibrium (equation 2.5.22), the relaxation times lying between 30 and 60 ns. The increase in the extrapolated low frequency values of α/f^2 with increasing helical content and the small excess absorption at low pH values suggest that the observed relaxation corresponds to a perturbation of the helix-coil equilibrium (Parker, Slutsky and Applegate, 1968). In contrast to this, the

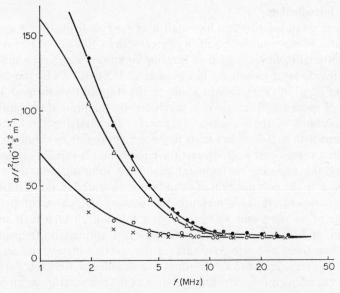

FIGURE 13.3.1. Ultrasonic absorption in 0·156 M poly-L-lysine in 0·6 M aqueous sodium chloride at 309·0 K: ●, pH 9·60; △, pH 9·22; ○, pH 7·20; ×, pH 5·10; —, single relaxation curves (after Parker, Applegate and Slutsky, 1966).

results of Burke, Hammes and Lewis (1965) for solutions of poly-L-glutamic acid in water–dioxan mixture show a distribution of relaxation times which implies that solvation of the polypeptide molecule is also involved. This is supported by the work of Hammes and Pace (1968) on aqueous solutions of glycine and of Hawley and Dunn (1969) on aqueous solutions of the uncharged polymer dextran.

A complete interpretation of the ultrasonic relaxation of polypeptide solutions is not yet available. Such studies will be of considerable importance in investigating the stability and denaturation of proteins (Kessler and Dunn, 1969) and the kinetics of enzyme–substrate binding.

CHAPTER 14

SOLUTION KINETICS

14.1 Introduction

Reactions in solution can have half-lives ranging from 10^{-10} s upwards. The rates of reactions with half-lives greater than 10 s can be measured by conventional techniques such as titration or spectroscopy. The time scale accessible to such techniques has been extended to 10^{-3} s by the development of rapid mixing techniques such as the stopped flow method, but the study of more rapid reactions is precluded by the finite time required to mix solutions of the reactants. Techniques for studying the kinetics of reactions with half-lives less than 10^{-3} s are applicable to solutions which contain a reaction at equilibrium: the equilibrium is rapidly displaced by altering the pressure or temperature of the solution, and the rate of approach to the new position of equilibrium measured (Caldin, 1964).

The temperature and pressure fluctuations, which accompany the passage of an ultrasonic wave through a solution, disturb such an equilibrium, and from the change in ultrasonic absorption with frequency the relaxation times and rate constants for the reaction may be deduced. The ultrasonic technique can cover about five decades of time (10^{-5}–10^{-10} s), and is complementary to techniques such as the temperature jump method (times greater than 10^{-6} s) or the electric impulse method (times longer than 10^{-7} s). Other techniques available for studying the rates of very fast reactions in solution include fluorescence quenching and electron spin resonance spectroscopy, but these are limited to electronically excited molecules and free radicals respectively. In the following Sections some reactions representative of those which have been studied by the ultrasonic method will be discussed.

14.2 Reactions in Solution

When the frequency of an ultrasonic longitudinal wave is low, equilibria which are present in solution are able to adjust themselves to the instantaneous values of temperature and pressure in the sound wave, and energy is abstracted from the sound wave. When the period of the sound wave is shorter than the relaxation time of an equilibrium however, the sound absorption is small. It is customary to discuss the ultrasonic absorption of solutions in terms of the excess absorption per wavelength $Q\lambda$:

$$Q\lambda = \Delta\alpha \, \lambda/n \tag{14.2.1}$$

Here $\Delta\alpha$ is the difference between the absorption coefficient of the solution and that of the solvent, λ is the ultrasonic wavelength, and n is the number of reactant molecules in unit volume of solvent.

Figure 14.2.1 shows the results of Tamm, Kurtze and Kaiser (1954) for the ultrasonic absorption $Q\lambda$ as a function of frequency in aqueous

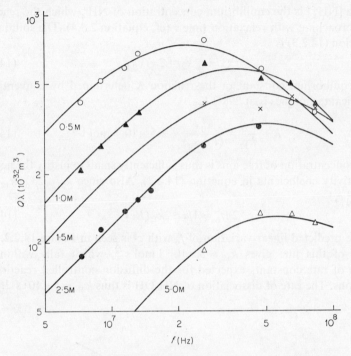

FIGURE 14.2.1. Frequency dependence of the excess absorption per molecule $Q\lambda$ at 293·2 K for aqueous solutions of ammonium hydroxide: —, single relaxation theory (equation 2.5.18); ○, 0·5 M; ▲, 1·0 M; ×, 1·5 M; ● 2·5 M; △, 5·0 M (after Tamm, Kurtze and Kaiser, 1954).

ammonium hydroxide at concentrations between 0·5 M and 5·0 M. The curves, which have a maximum at the characteristic frequency of the reaction $f_c = 1/2\pi\tau$, correspond to the dissociation of NH_4OH:

$$NH_4^+ + OH^- \underset{k_{21}}{\overset{k_{12}}{\rightleftharpoons}} NH_4OH \qquad (14.2.2)$$

$$\alpha_d c \qquad \alpha_d c \qquad\qquad (1-\alpha_d)c$$

α_d and c are the degree of dissociation and concentration of NH_4OH. The

9

rate equation of the reaction is

$$\frac{d[NH_4^+]}{dt} = k_{21}[NH_4OH] - k_{12}[NH_4^+][OH^-]$$

$$= ([NH_4^+] - [\overline{NH_4^+}])/\tau \qquad (14.2.3)$$

where $[\overline{NH_4^+}]$ is the equilibrium concentration of NH_4^+ which the reaction is approaching with relaxation time τ (cf. equation 2.3.4). The solution of equation (14.2.3) is

$$1/\tau = k_{21} + 2\alpha_d c k_{12} \qquad (14.2.4)$$

The equilibrium constant of the reaction K determined by conventional chemical techniques is at 298·2 K

$$K = \frac{k_{21}}{k_{12}} = \frac{\alpha_d^2 c}{(1 - \alpha_d)} = 1·5 \times 10^{-5} \text{ mol l}^{-1} \qquad (14.2.5)$$

The concentration of free ions is thus sufficiently small to justify the neglect of activity coefficients in equation (14.2.3). Also since $k_{21} \ll 2\alpha_d c k_{12}$ and $K \approx \alpha_d^2 c$,

$$2\pi f_c = 1/\tau \approx 2k_{12}(Kc)^{\frac{1}{2}} \qquad (14.2.6)$$

The predicted linear variation of f_c with $c^{\frac{1}{2}}$ is seen in Figure 14.2.2. The slope of this line gives $k_{12} = 3 \times 10^{10}$ l mol s^{-1}, which falls within the range of rate constants expected for the diffusion controlled reaction of two ions. The rate of dissociation of NH_4OH is thus $k_{21} = 5 \times 10^5$ s^{-1}.

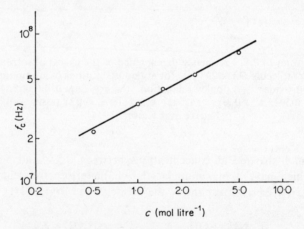

FIGURE 14.2.2. Double logarithmic plot of the variation of the characteristic relaxation frequency f_c with concentration c for aqueous solutions of ammonium hydroxide at 293·2 K (after Tamm, 1968).

Ultrasonic relaxation has also been observed in the hydration of SO_2 in aqueous solution, and in the hydrolysis of metaborate, phosphate and cyanide ions. In some cases the reactions have been studied at different temperatures, and the activation energy obtained from an Arrhenius plot of the logarithm of the rate constant against the reciprocal of the absolute temperature. Some representative results are given in Table 14.2.1.

TABLE 14.2.1. Rate constants for reactions in aqueous solution at 298 K

Reaction	k_{12} ($l\ mol^{-1}\ s^{-1}$)	k_{21} (s^{-1})	Reference*
$NH_4^+ + OH^- \rightleftharpoons NH_4OH$	$3\ \times 10^{10}$	$5\ \times 10^5$	1
$(C_2H_5)_3NH^+ + OH^- \rightleftharpoons (C_2H_5)_3N.H_2O$	$2 \cdot 0 \times 10^{10}$	$8\ \times 10^6$	2
$H_3O^+ + HSO_3^- \rightleftharpoons SO_2 + 2H_2O$	$2\ \times 10^8$	$3 \cdot 4 \times 10^6$	3
$HBO_2 + OH^- \rightleftharpoons BO_2^- + H_2O$	$2 \cdot 4 \times 10^8$	$4 \cdot 1 \times 10^3$	4
$HPO_4^{2-} + OH^- \rightleftharpoons PO_4^{3-} + H_2O$	$5 \cdot 0 \times 10^8$	$1 \cdot 1 \times 10^7$	5
$HCN + OH^- \rightleftharpoons CN^- + H_2O$	$3 \cdot 7 \times 10^9$	$5 \cdot 2 \times 10^4$	6

* References
1. Tamm, Kurtze and Kaiser (1954).
2. Brundage and Kustin (1970).
3. Eigen, Kustin and Maass (1961).
4. Yasunaga, Tatsumoto and Miura (1965).
5. Yasunaga, Tanoura and Miura (1965).
6. Stuehr and coworkers (1963).

The propagation of ultrasonic waves in carboxylic acids has been widely studied since the original observation of a relaxation process by Bazulin (1936). Figure 14.2.3 shows the various equilibria which may be involved.

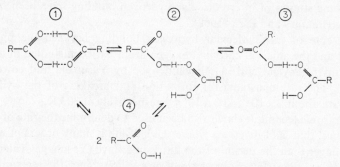

FIGURE 14.2.3. Possible equilibria in carboxylic acids.

Lamb and Pinkerton (1949) suggested that the combination of two acid molecules to form a dimer ($1 \rightleftharpoons 4$) is responsible for the observed relaxation. Piercy and Rao (1967c) point out that the activation energy $\Delta H^{\neq}$ of

34 kJ mol^{-1} for the relaxation in formic acid is similar to the value of the barrier to internal rotation $\Delta H_b^{\neq}$ in methyl and ethyl formate (Section 12.6) and to that in formic acid in the vapour phase (Lide, 1964). This implies that the equilibrium $2 \rightleftharpoons 3$ is being observed in carboxylic acids. Tatsumoto (1967) studied solutions of propionic acid. The characteristic frequency of the relaxation is independent of acid concentration, indicating a uni-unimolecular reaction such as $1 \rightleftharpoons 2$ in which a single hydrogen bond is broken. Since more than one relaxation is sometimes observed (Piercy and Lamb, 1956; Tatsumoto, 1967), several equilibria are probably involved, and an unambiguous explanation of ultrasonic propagation in carboxylic acids is not yet available.

The situation is much simpler in solutions of 2-pyridone which forms one of the most stable hydrogen-bonded dimers known (Figure 14.2.4). Hammes and Spivey (1966) observed a single relaxation time in solutions

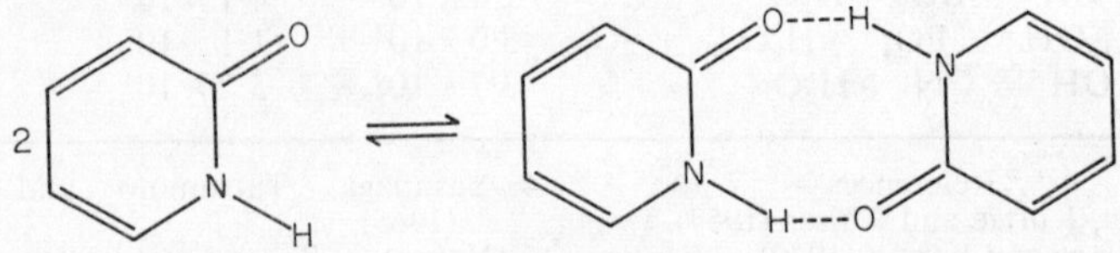

FIGURE 14.2.4. Dimerization of 2-pyridone.

of 2-pyridone in benzene and dioxan. The formation of the hydrogen bonds is diffusion controlled, and the bond strength may be deduced from the temperature dependence of the rate of dissociation of the dimer. Other ultrasonic studies of intermolecular hydrogen bonding have been carried out by Hammes and Park (1968).

The rate of intramolecular hydrogen bonding between the phenolic —OH and the adjacent carbonyl oxygen atom has been studied in methyl and ethyl salicylate and in salicylaldehyde by Yasunaga and coworkers (1969). The rate constants for formation and breaking of the hydrogen bond are about 7×10^5 and 3×10^7 s^{-1} respectively at 303 K.

Ultrasonic techniques have also been used to determine the rate of 'slow' reactions, that is of reactions where the half-life is many orders of magnitude longer than the period of the ultrasonic wave (Yasunaga, Tatsumoto and Miura, 1964; Pancholy and Saksena, 1967). Since the reactants and products in a reaction will have different sound velocities and absorption coefficients, the change in these quantities with time will depend on the rate of the reaction. Neither velocity nor absorption is a simple function of the composition of a mixture, however, and other methods (such as spectroscopy) for following the progress of the reaction are usually preferred.

14.3 Ion Association in 2-2 Electrolytes in Solution

Ultrasonic absorption in aqueous solutions of 2-2 electrolytes has been studied for many years because of the importance of this process in the propagation of underwater sound. Figure 14.3.1 shows the results of Tamm (1968) for 2 M $MnSO_4$ at 278 K. The linear increase of $Q\lambda$ with frequency at high frequencies is attributed to the contribution of the shear and volume viscosities of the solution to the absorption. The remainder of the curve

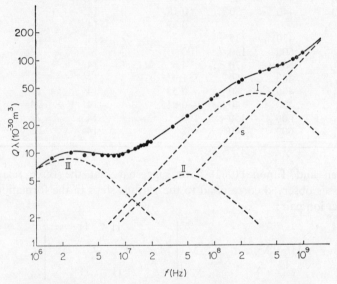

FIGURE 14.3.1. Frequency dependence of the excess absorption per molecule $Q\lambda$ at 278·2 K for 2 M $MnSO_4$ in water: •, experimental points; ---, curves showing contributions of three single relaxation processes (I, II, III) and solvent (s) to total absorption (—) (after Tamm, 1968).

can be described within experimental error by the superposition of three relaxation processes as shown in the Figure: the dependence of the characteristic frequencies of these processes on concentration is small.

Similar effects have been observed in a number of other solutions of 2-2 electrolytes by Kurtze and Tamm (1953), Tamm, Kurtze and Kaiser (1954), Fritsch and coworkers (1970) and Bechtler, Breitschwerdt and Tamm (1970). Table 14.3.1 lists the characteristic frequencies of the three processes. The lowest frequency process is strongly dependent on the cation although it is almost independent of the anion. The intermediate process is less strongly dependent on the cation, while the highest frequency process is almost independent of the cation.

TABLE 14.3.1. Relaxation frequencies and activation energies for ion pair formation in 0·5 M 2–2 electrolytes in water at 293·2 K (Bechtler, Breitschwerdt and Tamm, 1970)

| Salt | f_c (MHz) | | | $\Delta G_{21}^{\neq}$ | $\Delta G_{32}^{\neq}$ | $\Delta G_{43}^{\neq}$ |
	I	II	III		(kJ mol^{-1})	
BeSO$_4$	440	30	0·0002	14	24	54
MgSO$_4$	440	45	0·12	14	23	39
MgS$_2$O$_3$	440	30	0·18	14	24	38
MgCrO$_4$	—	—	0·17	14	—	38
VSO$_4$	430	35	—	14	24	—
CrSO$_4$	600	150	100	14	22	21
MnSO$_4$	500	70	4·6	14	21	30
FeSO$_4$	480	50	1·0	14	22	34
CoSO$_4$	480	60	0·52	14	21	35
NiSO$_4$	430	40	0·015	14	24	44
CuSO$_4$	600	150	100	14	22	21
ZnSO$_4$	600	140	58	14	21	21

Eigen and Tamm (1962) have suggested that the three relaxation processes observed correspond to the various steps in the formation of a contact ion pair:

$$M_{aq}^{2+} + A_{aq}^{2-} \underset{k_{21}}{\overset{k_{12}}{\rightleftharpoons}} \left[M^{2+} \begin{array}{c} H \\ \diagup \\ O \\ \diagdown \\ H \end{array} \begin{array}{c} H \\ \diagup \\ O \\ \diagdown \\ H \end{array} A^{2-} \right]_{aq}$$

$$(1) \hspace{4cm} (2) \hspace{3cm} (14.3.1)$$

$$\underset{k_{32}}{\overset{k_{23}}{\rightleftharpoons}} \left[M^{2+} \begin{array}{c} H \\ \diagup \\ O \\ \diagdown \\ H \end{array} A^{2-} \right]_{aq} \underset{k_{43}}{\overset{k_{34}}{\rightleftharpoons}} [M^{2+}A^{2-}]_{aq}$$

$$(3) \hspace{4cm} (4)$$

In reaction $1 \rightleftharpoons 2$ the solvated ions come together in a diffusion-controlled step to form an encounter pair. In reaction $2 \rightleftharpoons 3$ a water molecule is removed from the solvation sheath of the anion to give an ion pair in which the ions are separated by one water molecule, while in the slowest step $3 \rightleftharpoons 4$ the remaining water molecule is removed to form a contact ion pair.

The rate of exchange of water molecules between the first coordination sphere of the cation and the bulk of the solution [k_{34} in equation (14.3.1)] has been obtained by Swift and Connick (1962, 1964) and Chmelnick and

Fiat (1967) from nuclear magnetic relaxation measurements. k_{12} may be calculated from the theory of diffusion-controlled reactions, while the overall dissociation constant of the salt is known from electrical conductivity measurements. Hence a kinetic analysis can be used to obtain all the rate constants of equation (14.3.1) from the three observed relaxation frequencies. Table 14.3.2 gives as an example the equilibrium constants and

TABLE 14.3.2. Rate constants and equilibrium constants for ion pair formation in 0.5 M $MnSO_4$ in water at 293.2 K (Bechtler, Breitschwerdt and Tamm, 1970)

Reaction	k (s^{-1})	K (mol l^{-1})
2–1	1.0×10^9	0·025
3–2	1.2×10^9	0·77
4–3	2.6×10^7	1·1

forward rate constants for the steps in the formation of a contact ion pair in 0.5 M $MnSO_4$ at 293 K.

The activation energies for the various reaction steps have been obtained from the temperature dependence of the rate constants, and these are also given in Table 14.3.1. The activation energy $\Delta G_{43}^{\neq}$ for the substitution of a water molecule in the coordination sphere of the cation increases with decreasing ionic radius in Be^{2+}, Mg^{2+} and Zn^{2+}. In the first row transition metals, however, there are large variations in $\Delta G_{43}^{\neq}$ and in the relaxation frequency for reaction 3–4. Breitschwerdt (1967, 1968) has shown that these variations can be completely accounted for by ligand field stabilization.

14.4 Ion Association in Other Electrolyte Solutions

Ultrasonic relaxation has been observed in a number of other electrolyte solutions including Na_2SO_4, $La(NO_3)_3$ and $Al_2(SO_4)_3$ (Kurtze and Tamm, 1953; Tamm, Kurtze and Kaiser, 1954) and nickel acetate, $(NH_4)_2HPO_4$, $CuNH_4SO_4$ and $K_3Fe(CN)_6$ (Pancholy and Saksena, 1967). The high values of ultrasonic absorption in many other electrolytes suggest the existence of a relaxation process above the frequency range available experimentally. In some salts the ion association is likely to be complicated by hydrolysis reactions.

A study of the rare earth sulphates has been carried out by Purdie and Vincent (1967), Fay, Litchinsky and Purdie (1969) and Fay and Purdie

(1970), and some of their results for the absorption per molecule $Q\lambda$ are shown in Figure 14.4.1. Only one relaxation process is observed in each solution. Since the pH of the solutions used was less than 5, hydrolysis of the cations should not occur, and Purdie and Vincent (1967) interpreted

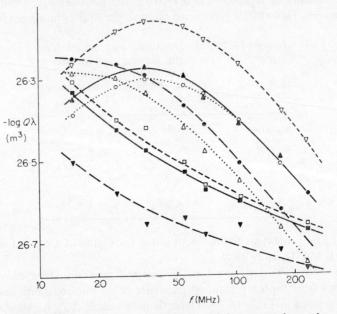

FIGURE 14.4.1. Frequency dependence of the excess absorption per molecule $Q\lambda$ at 298·2 K for 0·01 M aqueous solutions of lanthanide sulphates: ■, —, yttrium; □, ---, lanthanum; ●, --, cerium; ○, ···, samarium; ▽, ---, europium; ▲, —, gadolinium; △, ···, dysprosium; ▼, ---, ytterbium (after Purdie and Vincent, 1967).

the observed relaxation as step III of the Eigen–Tamm mechanism (equation 14.3.1):

$$M^{3+} \underset{H}{\overset{H}{\diagdown\diagup}} O \quad SO_4^{2-} \underset{k_{43}}{\overset{k_{34}}{\rightleftharpoons}} M^{3+}SO_4^{2-} \tag{14.4.1}$$

The rate of substitution of water in the inner coordination sphere of the cation is considerably higher than for divalent metal ions of the same size. The rate constant k_{34} for the elimination of water shows a maximum in the middle of the lanthanide series: this was attributed to an increase in the coordination number of the cation with increasing atomic number. Similar behaviour is shown by the rare earth acetates (Garza and Purdie, 1970): an additional low frequency relaxation is attributed to the removal of a second water molecule from the cation to give a bidentate acetate ligand.

In aqueous solutions of the alkali halides, ultrasonic relaxation is not usually observed because of the small interionic forces in a solvent of high dielectric constant (Plass and Kehl, 1968). One exception is concentrated aqueous CsF where a relaxation process with a characteristic frequency of about 20 MHz at room temperature has been observed. This was ascribed to an association (Tamm, 1968):

$$Cs^+ + F^- \; \rightleftharpoons \; CsF \tag{14.4.2}$$

In other alkali halides where the absorption α/f^2 is independent of frequency, α/f^2 at first decreases with increasing salt concentration and then increases at high concentrations. This behaviour can be attributed to the change in solvent structure induced by the ions (Breitschwerdt, Kistenmacher and Tamm, 1967; Montrose and Fritsch, 1970).

Kor, Rai and Awasthi (1969) find no evidence of a relaxation process in solutions of 1-1 electrolytes in a 50–50 mixture of water and ethylene glycol. In contrast to this, Blandamer and coworkers (1968a) observe a relaxation with a characteristic frequency in the region of 10 MHz for solutions of LiCl in a wide range of alcohols. They attribute this to the formation of ion pairs. The diffusion-controlled formation of ion pairs also leads to ultrasonic relaxation in solutions of the tetra-n-alkyl ammonium salts (Darbari and Petrucci, 1970), and of hydrogen chloride in solvents of low dielectric constant (Chan and Valleau, 1968).

14.5 Kinetics of Micelle Formation

Soaps and detergents consist of salts of long-chain aliphatic compounds. At low concentrations in water, such substances form a true solution in which the soap is dispersed ionically throughout the liquid. As the soap concentration is increased above a certain concentration called the 'critical micelle concentration', the soap ions aggregate into micelles in which the hydrophobic alkyl chains are situated in the centre of the micelle and the hydrophilic groups are on the surface. Such micelles are in dynamic equilibrium with each other and with 'free' soap ions in solution. The counterions (e.g. Na^+) are either adsorbed on the surface of the micelle or are in the bulk solution. Yasunaga, Oguri and Miura (1967) and Yasunaga, Fujii and Miura (1968) have studied the propagation of ultrasonics in a number of solutions of soaps and detergents. At concentrations below the critical micelle concentration, no relaxation is observed, but when micelles are formed by increasing the soap concentration or adding sodium chloride then ultrasonic relaxation occurs. The results may be described by the equation for a single relaxation process with characteristic frequency in the region of 10 MHz.

Since relaxation is observed only above the critical micelle concentration, the process can be attributed to a dissociation–recombination reaction between the micelles and the ions in solution. From an analysis of the various possible reactions, Yasunaga and coworkers concluded that the reaction being observed is

$$M'' \; \rightleftharpoons \; M' + nNa^+ \tag{14.5.1}$$

where M'' and M' are micelles containing different numbers of counterions. This suggestion is opposed by Zana and Lang (1968a, 1968b) who find that no change in the rate of reaction occurs for a number of different counterions. Instead, they consider that the relaxation process corresponds to the formation and dissociation of the micelle, that is, to an equilibrium between soap ions in the micelle and free soap ions in solution. This explanation agrees well with the results available.

14.6 Ultrasonic Vibration Potentials

A method of studying interactions in ionic solutions which is different from that described in Sections 14.3 and 14.4 was first suggested by Debye (1933). As an ultrasonic wave is passed through a liquid, a given small volume is exposed to alternate compressions and rarefactions. If this volume contains cations which have mobilities different from those of the anions, then the ultrasonic wave will induce local deviations from electrical neutrality in the solution. If two metal probes are inserted into the solution at a separation which is other than an integral multiple of the sound wavelength, an alternating electrical potential should be observed with frequency equal to that of the sound wave.

The forces acting on a given ion in a solution through which an ultrasonic wave is passing are the frictional force on the solvated ion as it moves, the relaxation force of the ion atmosphere, the electrophoretic force, the diffusion gradient and the pressure gradient. In moderately dilute aqueous solutions where the diffusion forces are negligible, the ultrasonic vibration potential ϕ of an electrolyte $M^{z+}X^{z-}$ can be shown to be (Zana and Yeager, 1967a)

$$\phi = 1 \cdot 55 \times 10^{-7} a_0 \left(\frac{t_+ W_+}{z_+} - \frac{t_- W_-}{z_-} \right) \tag{14.6.1}$$

where a_0 is the amplitude of the ultrasonic wave, t_+ and t_- are the transport numbers of cation and anion at infinite dilution, and W_+ and W_- are the apparent molar masses of the solvated cation and anion, that is, the difference between the molecular weight of the solvated ion and the mass of solvent displaced by one mole of ions.

Although experimental evidence for ionic vibration potentials was first reported by Yeager and coworkers (1949), a quantitative study of the effect was delayed by experimental difficulties: these arise in the measurement of an electrical signal of a few microvolts in the presence of many watts of electrical energy of the same frequency which are required to generate the ultrasonic signal. These difficulties were overcome in a careful investigation by Zana and Yeager (1967a). They used a pulsed system with an ultrasonic signal of frequency 220 or 500 kHz. In order to obtain a pulse length of 1 ms, the ultrasonic signal was transmitted through 2 m of water before entering the electrolyte cell through a plastic window. The voltage probes consisted of platinum wires sealed in glass. When spurious effects were eliminated, there was good agreement between the observed values of the ultrasonic vibration potential and the values predicted by equation (14.6.1). Table 14.6.1 contains some representative results in aqueous electrolyte solutions.

TABLE 14.6.1. Ultrasonic vibration potentials ϕ/a in aqueous electrolyte solutions at 220 kHz and 295 K (Zana and Yeager, 1967a)

Electrolyte	ϕ/a (μV s m^{-1})	
	10^{-3} M	10^{-1} M
HCl	20	60
LiCl	30	5
NaCl	115	60
KCl	190	170
RbCl	520	520
CsCl	810	810
MgCl$_2$	110	110
Mg(NO$_3$)$_2$	20	10

These values of the ultrasonic vibration potentials have been used to determine ionic partial molal volumes. These are defined as the difference between the actual volume of the solvated ion and the volume that the bound water would have in the absence of the ion, and provide an important measure of the strength of the interaction between ion and solvent. The ionic partial molal volume of the jth ion, $\bar{V}_j$, is

$$\bar{V}_j = (V_j)_h - (M_j)_W/\rho_s \qquad (14.6.2)$$

where $(V_j)_h$ is the molar volume of the solvated j ions, $(M_j)_W$ is the mass of water bound by one mole of j ions and ρ_s is the density of water. Now W_j in equation (14.6.1) is given by

$$W_j = (M_j)_h - (V_j)_h \rho_s$$
$$= M_j + (M_j)_W - (V_j)_h \rho_s \qquad (14.6.3)$$

where the molecular weight of the solvated ions $(M_j)_h$ equals the sum of the molecular weight of the ions M_j and the mass of bound water per mole of ions $(M_j)_W$. Combination of equations (14.6.2) and (14.6.3) gives the expression for the apparent molar mass of the jth ion:

$$W_j = M_j - \bar{V}_j \rho_s \qquad (14.6.4)$$

$(W_+ - W_-)$ for a given solution may be obtained from measurements of the ultrasonic vibration potential using equation (14.6.1), and this gives $(\bar{V}_+ - \bar{V}_-)$ from equation (14.6.4). $(\bar{V}_+ + \bar{V}_-)$ may be calculated from the partial molal volume of the electrolyte in solution (determined from density measurements), and so the individual values of $\bar{V}_+$ and $\bar{V}_-$ may be found. Typical values (Zana and Yeager, 1967a) of ionic partial molal volumes determined by the ultrasonic method are given in Table 14.6.2. These values represent the first absolute determination of ionic partial molal volumes, and there are some differences between these values and those estimated by other techniques.

TABLE 14.6.2. Ionic partial molal volumes $\bar{V}_j$ in aqueous electrolyte solutions at 295 K (Zana and Yeager, 1967a)

Ion	Li+	Na+	K+	Rb+	Cs+	Mg²⁺	Cl⁻	NO₃⁻
$\bar{V}_j$ (10^{-6} m³ mol⁻¹)	$-11\cdot2$	$-7\cdot4$	$3\cdot4$	$9\cdot0$	$15\cdot5$	$-35\cdot6$	$23\cdot7$	$34\cdot8$

Ultrasonic vibration potentials have also been studied in colloidal systems (Yeager, Dietrick and Hovorka, 1953). Here the large mass difference between the colloidal particle and the counterion produces vibration potentials up to 1 V s m⁻¹. Similar behaviour is shown by polyelectrolytes if the polyion is tightly coiled. On the other hand, if the polyion is in an extended conformation the motions of the chain segments are to some extent independent of each other and the polyelectrolyte vibration potential may be calculated by considering the molecule to be composed of a chain of beads in which each bead develops an ionic vibration potential (Zana and Yeager, 1967b). There is quantitative agreement between the observed polyelectrolyte vibration potential and the value calculated from the above model. Such measurements give information on the bonding between the counterions and the polyion and on the effective volumes of the monomer units in the polymer beads.

14.7 Ultrasonic Absorption in Non-electrolyte Solutions

The absorption coefficient for ultrasonic waves in many polar liquids such as alcohols or amines is similar to that in water. When such a solute is added to water, however, the absorption coefficient of the solution increases greatly and attains a maximum value at a concentration of about 10% solute: the maximum value of α/f^2 is typically ten times that of the solute or solvent. The frequency dependence of the absorption indicates at least two relaxation processes in most solutions.

At very low concentrations of amines in water, the excess absorption may be attributed to an acid-base equilibrium (McKellar and Andreae, 1962; Blandamer and coworkers, 1967):

$$R_3N + H_2O \; \rightleftharpoons \; R_3NH^+ + OH^- \qquad (14.7.1)$$

At higher concentrations of amine, this reaction would be too rapid to be observed within the conventional frequency range. Blandamer and coworkers (1969) then ascribe the relaxation to an equilibrium involving different clathrate hydrates. Their model assumes that a range of hydrate structures occurs in solution alongside structures where two amine molecules occupy the same solvent cavity. The equilibria between the various structures could give rise to the observed relaxations. A similar model has been applied to the excess absorption in aqueous solutions of alcohols (Blandamer and coworkers, 1968b, c) while Solovyev and coworkers (1968) attribute the maximum in α/f^2 at low concentrations of ethanol in the ethyl halides to the disruption of the local order of the solvent by the molecules of the solute. In general the causes of the ultrasonic absorption in non-electrolyte solutions are poorly understood.

CHAPTER 15

PROPAGATION OF ULTRASONIC WAVES IN SOLIDS

15.1 Introduction

The propagation of ultrasonic longitudinal and shear waves in solids is an important technique for studying the solid state. Ultrasonic waves with frequencies greater than 1 MHz are generally used, and the methods of measuring the velocity and absorption coefficient of such waves have been discussed in Chapter 3. The present Chapter gives a brief outline of some of the phenomena which may be studied in this way: fuller accounts can be found in Mason (1958, 1964) and in Truell, Elbaum and Chick (1969).

The adiabatic elastic constants of crystals may be calculated from the velocity of ultrasonic waves. In the case of a longitudinal wave propagating along the [100] direction of a cubic crystal, the ultrasonic velocity U_1 is related to the second-order elastic constant c_{11} (Young's modulus) by the expression (cf. equation 2.1.12)

$$U_1^2 = c_{11}/\rho \tag{15.1.1}$$

where ρ is the crystal density (Kittel, 1966). Similarly the longitudinal velocity U_2 along the [110] direction of a cubic crystal is related to the three second-order elastic constants c_{11}, c_{12} and c_{44}:

$$U_2^2 = (c_{11} + c_{12} + 2c_{44})/2\rho \tag{15.1.2}$$

The rigidity modulus c_{44} of a cubic crystal may be found from the velocity of a shear wave U_s along the [100] direction:

$$U_s^2 = c_{44}/\rho \tag{15.1.3}$$

Thus the three elastic constants of a cubic crystal may be determined, and also the bulk modulus K from the relation

$$K = \tfrac{1}{3}(c_{11} + 2c_{12}) \tag{15.1.4}$$

Similar relations exist for the other crystal systems (Brugger, 1965).

The second-order elastic constants arise when Hooke's law is applicable to the deformation of a crystalline solid. Deviations from Hooke's law, when the lattice potential energy no longer has a harmonic form, are characterized by the third-order elastic constants. These may be determined from the changes in ultrasonic velocity that occur when a crystal is subjected to a steady stress (Brugger, 1964), or from the distortion of an ultrasonic wave of large amplitude during its passage through the solid (Gauster and Breazale, 1968).

Ultrasonic absorption in solids is generally less than in liquids and much less than in gases. There is a contribution from the thermal conductivity of the solid (cf. equation 2.2.5), although this is small except in metals. The remainder of this Chapter is devoted to a survey of the main causes of ultrasonic absorption in solids.

15.2 Thermoelastic Effects

A real solid contains a large number of defects ranging in size from grain boundaries to random absences of ions from lattice sites. The thermoelastic effect is found in crystalline materials where the different grains have random orientations with respect to a longitudinal wave. The elastic constants of a crystalline grain depend on its orientation (Section 15.1), so that a given macroscopic stress will produce different strains in different grains. This gives rise to random local temperature variations superimposed on the oscillatory temperature variations of the longitudinal wave. The resulting flow of heat between adjacent grains will lead to the dissipation of acoustic energy and to ultrasonic absorption (Zener, 1938, 1940).

Thermoelastic relaxation usually occurs at frequencies below 1 MHz. The contribution to the total ultrasonic absorption is small, and this effect is not observed with shear waves.

15.3 Scattering

A polycrystalline solid contains a large number of grains which are elastically anisotropic and are oriented randomly. At each grain boundary there is a discontinuity in the elastic modulus and in the impedance presented to an elastic wave. The resulting reflexions of the elastic waves will lead to their attenuation, and the value of this attenuation will depend on the elastic anisotropy, on the diameter of the grains and on the sound wavelength. The quantitative calculation of the attenuation is complicated and has only been achieved in certain idealized cases (Truell, Elbaum and Chick, 1969).

Two situations arise depending on the relative sizes of the sound wavelength λ and the grain diameter d. When $\lambda \ll d$, the sound wave may be considered to travel through an individual grain without scattering loss. When it reaches the grain boundary, most of the sound energy will undergo refraction into the next grain, but part will be reflected back into the grain. The total attenuation in the medium will be determined chiefly by the number of times a wave meets a grain boundary and by the magnitude of the impedance discontinuity there. Hence, for a given material, the sound absorption will be independent of frequency and inversely proportional to the diameter of the grains (Mason and McSkimin, 1948).

The scattering of sound by particles which are much smaller than the sound wavelength was considered by Rayleigh (1877). The attenuation is proportional to the fourth power of the frequency and to the cube of the diameter of the scattering particle. In polycrystalline materials when $\lambda \ll d$, the attenuation α_R caused by Rayleigh scattering is

$$\alpha_R = R_S d^3 f^4 / U^3 \qquad (15.3.1)$$

where U is the sound velocity, and the elastic anisotropy factor R_S is constant for a given material and for a given type of wave motion (longitudinal or transverse).

Point defects (interstitial atoms, lattice vacancies or impurity atoms) produce local variations of the elastic constants of a solid which may lead to scattering of elastic waves. This effect is negligible at all but the highest frequencies (Splitt, 1969).

15.4 Dislocation Damping of Elastic Waves

The rigidity moduli observed in real crystals are usually several orders of magnitude less than the values expected for a perfect single crystal. The difference is caused by edge and screw dislocations in the real crystal. Figure 15.4.1 shows a section through a crystal where an edge dislocation

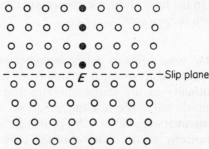

FIGURE 15.4.1. Pictorial representation of an edge dislocation E caused by the plane of atoms ●. The diagram represents a section through a crystal at right angles to the slip plane (---).

E arises because a plane of atoms ● is displaced from its correct position in the lattice. The extension of the dislocation at right angles to the plane of the paper is called the length of the dislocation line. The relatively easy movement of the dislocation along the slip plane permits relative motion of the parts of the crystal on either side of the slip plane. In a screw dislocation, the dislocation line is parallel to the slip direction (Cottrell, 1953).

The region around a dislocation is considerably strained. To relieve this strain, dislocations are joined in a network and are anchored or pinned at the network intersections. Point defects such as impurity atoms or interstitial atoms can also relieve the strain and act as somewhat weaker pinning points (separation L_C) between the network pinning points (separation L_N). When a stress is applied to the crystal, the dislocation loops of leegth L_C bow out until with increasing stress breakaway occurs and the loop length increases rapidly from L_C to L_N.

There are four main processes which lead to dislocation damping of an ultrasonic wave. These have been studied extensively and are reviewed by Truell, Elbaum and Chick (1969). In the megahertz frequency range, a frequency-dependent resonance loss occurs (Granato and Lücke, 1956). A dislocation loop of length L_C which is pinned at its two ends acts as a stretched string which can vibrate back and forth when excited by the alternating stress of an elastic wave. Some of the vibrational energy of the dislocation is degraded into heat by frictional forces in the crystal, and the elastic wave is attenuated.

If the applied stress exceeds a certain value, the loops of length L_C may break away from the weaker point defect pinning to become loops of length L_N (Granato and Lücke, 1956). These loops bow out during the loading part of the stress cycle while during the unloading cycle the loops collapse to become pinned again at the impurity sites. The loss of acoustic energy per cycle in this process is independent of frequency (hysteresis loss) but depends on the amplitude of the strain. At very high frequencies this effect is not usually observed because the strain is too small.

Thermal effects also contribute to the dislocation damping of elastic waves. If a dislocation is fixed at both its ends and has a point defect to act as an intermediate pinning point, a detailed consideration of the interaction between defect and dislocation shows that three configurations of the dislocation can exist depending on the stress involved (Teutonico, 1958). For low stresses only one potential minimum exists, but above a certain stress there are two potential wells and the dislocation can surmount the potential barrier between these wells by thermal activation. Such a two-state equilibrium leads to ultrasonic absorption in the kilohertz frequency range.

The fourth mechanism which contributes to the absorption of elastic waves is Bordoni relaxation (Bordoni, 1954, 1960). This effect is observed in crystals which contain a large number of dislocations, and arises from the local motion of part of a dislocation loop of length less than L_C. Because of the thermal energy of the crystal, the dislocation line does not always follow the path of lowest energy through the crystal and small lengths of the dislocation line deviate to positions of higher energy. An elastic wave disturbs this equilibrium, and the absorption per wavelength μ goes through a maximum at the characteristic frequency for the process.

Bordoni relaxation is also a two-state process, and the relaxation frequency varies exponentially with temperature (Seeger, Donth and Pfaff, 1957).

15.5 Absorption of Sound Waves in Solids by their Interaction with Phonons and Electrons

The total thermal motion of the atoms in a solid may be represented as the resultant of a large number of longitudinal and transverse elastic waves, i.e. thermal phonons. Such phonons can exchange energy with each other because of the anharmonic nature of the interatomic potential (Kittel, 1966). When an elastic wave is passed through a solid, the equilibrium distribution of thermal phonons is disturbed since by modulating the interatomic separation the sound wave modulates the phonon frequencies. The perturbed phonon distribution relaxes back to its equilibrium distribution by energy exchange among the thermal phonons, and this leads to a loss of energy from the sound wave at frequencies above 1 GHz. This absorption mechanism is important at temperatures above 100 K (Bömmel and Dransfeld, 1960). It makes a large contribution to the ultrasonic absorption at temperatures close to the melting point of a solid (Saunders, Pace and Alper, 1967).

In molecular crystals the lattice phonons are in equilibrium with the internal molecular vibrations and rotations. If the coupling is weak, the transfer of energy between the phonons and the internal modes may lead to a relaxational absorption similar to that discussed for gases in Chapter 4. The effect was observed by Liebermann (1959) in single crystals of benzene, and is present in other substances including naphthalene (Yun and Beyer, 1964) and carbon tetrachloride (Rasmussen, 1967).

The transfer of energy between the conduction electrons in a metal and the phonons is the main cause of ultrasonic absorption at temperatures below about 20 K. An elastic wave displaces the positively charged metal ions from their equilibrium positions and an oscillating electric field is generated in the metal. The electrons move under the influence of this field and at the same time collide with thermal phonons and impurities in the metal. As a result the distribution of electron velocities is not in equilibrium with the local electric field. The ensuing relaxation to equilibrium leads to dissipation of the acoustic energy and hence to sound absorption (Pippard, 1955). In a superconducting metal, on the other hand, there is negligible interaction between electrons and sound waves and the sound attenuation is small (Truell, Elbaum and Chick, 1969).

15.6 Sound Propagation in Ferromagnetic Materials

Ferromagnetic solids contain small regions or domains of typically 0·1 mm diameter in which the electron spins and magnetic moments are all aligned in a particular direction. Consider two adjacent domains A and B (Figure 15.6.1) in which the directions of magnetization are at right angles. Because of the magnetostrictive effect (change of length on magnetization) there is an increase in the length of the domains along the direction of

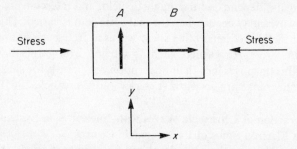

FIGURE 15.6.1. Domains in a ferromagnetic material under stress. The heavy arrows indicate the direction of magnetization in the domain.

polarization with an accompanying decrease in length in the direction at right angles to this. Thus domain A has expanded along the y direction and B along the x direction, while A has contracted along the x direction and B along the y direction. This leads to a displacement of the boundary or Bloch wall between the two domains.

If a compression is now applied along the x axis, the strain in this direction is reduced by the rotation of some of the magnetic moments in B to be parallel to those of A. Thus A grows at the expense of B and the Bloch wall moves along the line of the applied stress. When a cyclic stress is applied, the Bloch wall also executes a cyclic motion. This displacement does not vary linearly with the applied stress, however, but follows a hysteresis loop so that absorption of an ultrasonic longitudinal wave occurs. This absorption increases with strain amplitude and also with frequency, although at very high frequencies it decreases when the domain walls become unable to follow the rapid alterations of the applied stress. Similar effects are observed with shear waves.

A vibrating Bloch wall is equivalent to a periodically varying magnetic field. Such a field induces eddy currents within the domain so that some of the elastic energy is dissipated in Joule heating. These effects are important in the frequency range 100 kHz–100 MHz (Mason, 1958).

In a ferromagnetic domain where all the electron spins are aligned parallel, a low-lying excited state involves the precession of the spins about the direction of magnetization with successive spins advanced in phase by a constant amount. Such a motion is known as a 'spin wave' or 'magnon'. There exists a magnetostrictive coupling between the lattice vibrations and the spin waves. This means that shear waves propagating along the direction of an applied magnetic field will not be pure phonons but will consist of hybrid magnetoelastic waves in which the variations of lattice position and magnetization are combined. In a given solid, the exact type of motion that propagates depends on the magnetic field, and interconversion of spin and shear waves may occur when the magnetic field is varied. This provides an efficient method of generating shear waves at frequencies above 1 GHz, and certain rare earth garnets such as yttrium iron garnet have proved useful for this purpose since their absorption coefficients for shear waves in this frequency range are very low (LeCraw and Comstock, 1965).

15.7   Interaction of Ultrasonic Waves with Nuclear Spin Systems and with Electron Spins of Paramagnetic Centres

When a nuclear spin system is placed in a magnetic field, the energy levels of the nucleus are split. In conventional nuclear magnetic resonance experiments, the transitions between the nuclear energy levels are studied by observing the absorption of electromagnetic radiation of frequency corresponding to the energy difference between the nuclear energy levels: this is typically between 10 and 100 MHz for easily accessible magnetic fields. Ultrasonic waves can also excite nuclear spins to a higher energy level. The modulation of the atomic positions during the passage of an elastic wave leads to a variation in the magnetic field at the nucleus under investigation. If the frequency of this variation corresponds to the energy difference between the nuclear energy levels, the nucleus will be excited and the sound wave attenuated.

The small population difference between the nuclear energy levels means that the net ultrasonic absorption is small and special methods of measurement must be used (Melcher, Bolef and Merry, 1968). It is often more convenient to measure the decrease of the conventional n.m.r. signal as the populations of the nuclear energy levels are equalized by a high intensity ultrasonic signal.

Nuclear acoustic resonance has found applications in cases where the usual technique of magnetic field modulation presents difficulties. For example, in metals the skin depth may be much smaller than the sample dimensions so that the penetration and uniformity of the r.f. magnetic field are decreased. The acoustic technique also has advantages where the nuclear spin energy levels are determined by internal fields, and where several competing spin relaxation processes exist. Nuclear acoustic resonance has been reviewed by Al'tshuler, Kochelaev and Leushin (1962) and Bolef (1966).

An ultrasonic wave (ca. 10 GHz) can also induce transitions between the electronic energy levels of a paramagnetic ion when these are split by the application of an external magnetic field. If the ultrasonic frequency and the magnetic field are chosen appropriately, the modulation of the crystalline electric field at the ion induces a transfer of energy from the lattice vibrations to the electron. This causes the electron to be excited to a higher energy state by reversal of its spin. Such phonon-induced electron spin resonance is the converse of the spin lattice relaxation process which leads to line broadening in conventional photon-induced e.s.r. Hence acoustic e.s.r. is readily observed in those cases where line broadening interferes with the detection of photon-induced resonance. Both the electronic energy levels and the spin lattice coupling of a paramagnetic ion may be investigated by the acoustic technique (Taylor and Pointon, 1969). Since the selection rules for acoustic e.s.r. are less rigid than those for photon-induced e.s.r., it is possible to observe transitions between levels which cannot be detected with conventional e.s.r.

15.8 Ultrasonic Propagation in Semiconductors

In a semiconductor the number density of charge carriers—electrons and holes—is much smaller than the density of conduction electrons in a metal. As a result the ultrasonic attenuation caused by electrons and holes in a semiconductor is much less than that by electrons in a metal (Section 15.5). The semiconductor attenuation can be varied by altering the number of charge carriers by illumination (Nine and Truell, 1961) or by doping (Mason and Bateman, 1963). The existence of two types of charge carrier in semiconductors permits the observation of the acousto-electric effect in which a d.c. electric field is set up in a specimen through which a sound wave is passing. The momentum transferred from the sound wave to the conduction electrons (Section 15.5) produces a drift current in the direction of propagation of the ultrasonic wave, and opposite faces of the specimen become oppositely charged (Weinreich, Sanders and White, 1959).

When the wavelength of a sound wave is greater than the mean free path of the conduction electrons in a piezoelectric semiconductor, the electrons bunch in the region of low electrostatic potential in the sound wave. These electrons are carried along by a progressive sound wave which is consequently attenuated. If by the application of an external d.c. electric field the electrons are given a drift velocity greater than the ultrasonic velocity, the electrons will tend to pull the sound wave along, and amplification of the sound wave results (Hutson, McFee and White, 1961). This effect is observed with both longitudinal and transverse waves (McFee, 1966).

REFERENCES

Abraham, R. J. and J. A. Pople (1960). *Mol. Phys.*, **3**, 609.

Adler, F. T., W. M. Sawyer and J. D. Ferry (1949). *J. Appl. Phys.*, **20**, 1036.

Alfrey, T. (1948). *Mechanical Behaviour of High Polymers*, Interscience, New York.

Allis, J. W. and J. D. Ferry (1965). *J. Am. Chem. Soc.*, **87**, 4681.

Altberg, W. (1907). *Ann. Physik*, **23**, 267.

Alterman, E. B. and D. J. Wilson (1965). *J. Chem. Phys.*, **42**, 1957.

Al'tshuler, S. A., B. I. Kochelaev and A. M. Leushin (1962). *Sov. Phys. Uspekhi*, **4**, 880. [(1961). *Uspekhi Fiz. Nauk*, **75**, 459.]

Amme, R. and S. Legvold (1959). *J. Chem. Phys.*, **30**, 163.

Anantaraman, A. V., A. B. Walters, P. D. Edmonds and C. J. Pings (1966). *J. Chem. Phys.*, **44**, 2651.

Anderson, G. P. and B. B. Chick (1969). *J. Acoust. Soc. Am.*, **45**, 1343.

Andrade, E. N. da C. (1934). *Phil. Mag.* **17**, 497, 698.

Andreae, J. H., E. L. Heasell and J. Lamb (1956). *Proc. Phys. Soc. B*, **69**, 625.

Andreae, J. H. and J. Lamb (1956). *Proc. Phys. Soc. B*, **69**, 814.

Andreae, J. H., R. Bass, E. L. Heasell and J. Lamb (1958). *Acustica*, **8**, 131.

Andreae, J. H., P. L. Joyce and R. J. Oliver (1960). *Proc. Phys. Soc.*, **75**, 82.

Andreae, J. H. and P. L. Joyce (1962). *Brit. J. Appl. Phys.*, **13**, 462.

Arnold, J. W., J. C. McCoubrey and A. R. Ubbelohde (1958). *Proc. Roy. Soc. A*, **248**, 445.

Ashworth, J. N. and J. D. Ferry (1949). *J. Am. Chem. Soc.*, **71**, 622.

Baccaredda, M., E. Butta, V. Frosini and P. L. Magagnini (1966). *J. Polymer Sci. A2*, **4**, 789.

Bader, M. (1968). *Acustica*, **20**, 41.

Bagatskii, M. I., A. V. Voronel and V. G. Gusak (1963). *Sov. Phys. J.E.T.P.*, **16**, 517. [(1962). *Zh. Eksp. Teor. Fiz.*, **43**, 728.]

Bailey, J. and A. M. North (1968). *Trans. Faraday Soc.*, **64**, 1499.

Baker, C. B. and N. de Haas (1964). *Phys. Fluids*, **7**, 1400.

Baker, C. E. and R. S. Brokaw (1964). *J. Chem. Phys.*, **40**, 1523.

Baker, C. E. and R. S. Brokaw (1965). *J. Chem. Phys.*, **43**, 3519.

Baker, C. E. (1967). *J. Chem. Phys.*, **46**, 2846.

Baker, W. O., W. P. Mason and J. H. Heiss (1952). *J. Polymer Sci.*, **8**, 129.

Baranskii, K. N. (1957). *Sov. Phys. Doklady*, **2**, 237. [*Doklady Akad. Nauk S.S.S.R.*, **114**, 517.]

Barlow, A. J. and J. Lamb (1959). *Proc. Roy. Soc. A*, **253**, 52.

Barlow, A. J., G. Harrison, J. Richter, H. Seguin and J. Lamb (1961). *Lab. Pract.*, **10**, 786.

Barlow, A. J., G. Harrison and J. Lamb (1964). *Proc. Roy. Soc. A*, **282**, 228.

Barlow, A. J., J. Lamb and A. J. Matheson (1966). *Proc. Roy. Soc. A*, **292**, 322.

Barlow, A. J., J. Lamb, A. J. Matheson and J. Richter (1966). *Chem. Soc. Spec. Publ.*, **20**, 203.

Barlow, A. J. and S. Subramanian (1966). *Brit. J. Appl. Phys.*, **17**, 1201.

Barlow, A. J. and E. Yazgan (1966). *Brit. J. Appl. Phys.*, **17**, 807.

Barlow, A. J., R. A. Dickie and J. Lamb (1967). *Proc. Roy. Soc. A*, **300**, 356.

Barlow, A. J., A. Erginsav and J. Lamb (1967). *Proc. Roy. Soc. A*, **298**, 481.

Barlow, A. J., J. Lamb, A. J. Matheson, P. R. K. L. Padmini and J. Richter (1967). *Proc. Roy. Soc. A*, **298**, 467.

Barlow, A. J., M. A. Day, G. Harrison, J. Lamb and S. Subramanian (1969). *Proc. Roy. Soc. A*, **309**, 497.

Barlow, A. J., A. Erginsav and J. Lamb (1969). *Proc. Roy. Soc. A*, **309**, 473.

Barlow, A. J., J. Lamb and N. S. Tasköprülü (1969). *J. Acoust. Soc. Am.*, **46**, 569.

Barocchi, F., M. Mancini and R. Vallauri (1968). *J. Chem. Phys.*, **49**, 1935.

Barua, A. K., A. Manna and P. Mukhopadhyay (1968). *J. Chem. Phys.*, **49**, 2422.

Basco, N., A. B. Callear and R. G. W. Norrish (1961). *Proc. Roy. Soc. A*, **260**, 459.

Bass, R. and J. Lamb (1957). *Proc. Roy. Soc. A*, **243**, 94.

Bass, R. (1958). *J. Acoust. Soc. Am.*, **30**, 602.

Bass, R. and J. Lamb (1958). *Proc. Roy. Soc. A*, **247**, 168.

Bateman, T. B. (1967). *J. Acoust. Soc. Am.*, **41**, 1011.

Bauer, H. J. and K. F. Sahm (1965). *J. Chem. Phys.*, **42**, 3400.

Bauer, H. J. and H. Kosche (1966). *Acustica*, **17**, 96.

Bauer, H. J. and R. Schotter (1969). *J. Chem. Phys.*, **51**, 3261.

Bazulin, P. (1936). *Compt. Rend. Acad. Sci. U.R.S.S.*, **3**, 285.

Bechtler, A., K. G. Breitschwerdt and K. Tamm (1970). *J. Chem. Phys.*, **52**, 2975.

Beckerle, J. C. (1953). *J. Chem. Phys.*, **21**, 2034.

Bell, J. F. W. (1950). *Proc. Phys. Soc. B*, **63**, 958.

Bell, J. F. W. (1968). *Ultrasonics*, **6**, 11.

Benedek, G. B., J. B. Lastovka, K. Fritsch and T. Greytak (1964). *J. Opt. Soc. Am.*, **54**, 1284.

Bentley, J. B., K. B. Everard, R. J. B. Marsden and L. E. Sutton (1949). *J. Chem. Soc.*, 2957.

Berend, G. C. and S. W. Benson (1969). *J. Chem. Phys.*, **51**, 1480.

Berge, J. W., P. R. Saunders and J. D. Ferry (1959). *J. Colloid. Sci.*, **14**, 135.

Bergmann, L. (1949). *Der Ultraschall*, S. Hirzel Verlag, Zurich.

Bernal, J. D. (1964). *Proc. Roy. Soc. A*, **280**, 299.

Bhatia, A. B. (1967). *Ultrasonic Absorption*, Clarendon, Oxford.

Bhatia, A. B. and E. Tong (1968). *Phys. Rev.*, **173**, 231.

Biquard, P. (1931). *Compt. Rend.*, **193**, 226.

Biquard, P. (1933). *Compt. Rend.*, **196**, 257.

Biquard, P. (1936). *Ann. Phys.*, **6**, 195.

Biquard, P. and C. Ahier (1943). *Cahiers Phys.*, **15**, 21.

Birnboim, M. H. and J. D. Ferry (1961). *J. Appl. Phys.*, **32**, 2305.

Blandamer, M. J., D. E. Clarke, N. J. Hidden and M. C. R. Symons (1967). *Trans. Faraday Soc.*, **63**, 66.

Blandamer, M. J., D. E. Clarke, N. J. Hidden and M. C. R. Symons (1968a). *Trans. Faraday Soc.*, **64**, 2683.

Blandamer, M. J., D. E. Clarke, N. J. Hidden and M. C. R. Symons (1968b). *Trans. Faraday Soc.*, **64**, 2691.

Blandamer, M. J., N. J. Hidden, M. C. R. Symons and N. C. Treloar (1968c). *Trans. Faraday Soc.*, **64**, 3242.

Blandamer, M. J., N. J. Hidden, M. C. R. Symons and N. C. Treloar (1969). *Trans. Faraday Soc.*, **65**, 2663.

Blitz, J. (1967). *Fundamentals of Ultrasonics*, Butterworths, London.

Blizard, R. B. (1951). *J. Appl. Phys.*, **22**, 730.

Blythe, A. R., T. L. Cottrell and A. W. Read (1961). *Trans. Faraday Soc.*, **57**, 935.

Blythe, A. R., A. E. Grosser and R. B. Bernstein (1964). *J. Chem. Phys.*, **41**, 1917.

Blythe, A. R., T. L. Cottrell and M. A. Day (1965). *Acustica*, **16**, 118.

Boade, R. R. and S. Legvold (1965). *J. Chem. Phys.*, **42**, 569.

Bolef, D. I. (1966). In W. P. Mason (Ed.), *Physical Acoustics*, Vol. 4A, Academic Press, New York.

Boltzmann, L. (1876). *Pogg. Ann. Phys.*, **7**, 108.

Boltzmann, L. (1895). *Nature*, **51**, 413.

Bömmel, H. E. and K. Dransfeld (1960). *Phys. Rev.*, **117**, 1245.

Bordoni, P. G. (1954). *J. Acoust. Soc. Am.*, **26**, 495.

Bordoni, P. G. (1960). *Nuov. Cim. Suppl.*, **17**, 43.

Botch, W. and M. Fixman (1965). *J. Chem. Phys.*, **42**, 199.

Bradley, J. N. (1962). *Shock Waves in Chemistry and Physics*, Methuen, London.

Brau, C. A. and R. M. Jonkman (1970). *J. Chem. Phys.*, **52**, 477.

Breazale, M. A. (1962). *J. Chem. Phys.*, **36**, 2530.

Breitschwerdt, K. G. (1967). *Chem. Phys. Lett.*, **1**, 481.

Breitschwerdt, K. G., H. Kistenmacher and K. Tamm (1967). *Phys. Lett. A*, **24**, 550.

Breitschwerdt, K. G. (1968). *Ber. Bunsen. Phys. Chem.*, **72**, 1046.

Breshears, W. D. and P. F. Bird (1969a). *J. Chem. Phys.*, **50**, 333.

Breshears, W. D. and P. F. Bird (1969b). *J. Chem. Phys.*, **51**, 3660.

Breshears, W. D. and P. F. Bird (1970). *J. Chem. Phys.*, **52**, 999.

Brillouin, L. (1922). *Ann. Phys.*, **17**, 88.

Brout, R. (1954). *J. Chem. Phys.*, **22**, 934.

Brown, A. E. and E. G. Richardson (1959). *Phil. Mag.*, **4**, 705.

Brugger, K. and W. P. Mason (1961). *Phys. Rev. Lett.*, **7**, 270.

Brugger, K. (1964). *Phys. Rev.*, **133**, A1611.

Brugger, K. (1965). *J. Appl. Phys.*, **36**, 759.

Brundage, R. S. and K. Kustin (1970). *J. Phys. Chem.*, **74**, 672.

Bueche, F. (1954). *J. Chem. Phys.*, **22**, 603.

Bueche, F., C. J. Coven and B. J. Kinzig (1963). *J. Chem. Phys.*, **39**, 128.

Bugl, P. and S. Fujita (1969). *J. Chem. Phys.*, **50**, 3137.

Burke, J. J., G. G. Hammes and T. B. Lewis (1965). *J. Chem. Phys.*, **42**, 3520.

Burnett, D. (1935). *Proc. London Math. Soc.*, **39**, 385; **40**, 382.

Burundukov, K. M. and V. F. Yakovlev (1968). *Russ. J. Phys. Chem.*, **42**, 1141. [*Zh. Fiz. Khim.*, **42**, 2149.]

Burundukov, K. M. and V. F. Yakovlev (1969). *Sov. Phys. Acoust.*, **15**, 254. [*Akust. Zh.*, **15**, 295.]

Butcher, S. S. and E. B. Wilson (1964). *J. Chem. Phys.*, **40**, 1671.

Cady, W. G. (1946). *Piezoelectricity*, McGraw–Hill, New York.

Caldin, E. F. (1964). *Fast Reactions in Solution*, Blackwell, Oxford.

Caloin, M. and S. Candau (1969). *Compt. Rend. B*, **269**, 423.

Calvert, J. B. and R. C. Amme (1966). *J. Chem. Phys.*, **45**, 4710.

Capps, W., P. B. Macedo, B. O'Meara and T. A. Litovitz (1966). *J. Chem. Phys.*, **45**, 3431.

Carnevale, E. H. and T. A. Litovitz (1955). *J. Acoust. Soc. Am.*, **27**, 547.

Carome, E. F., F. A. Gutowski and D. E. Schuele (1957). *Am. J. Phys.*, **25**, 556.

Carpenter, M. R., D. B. Davies and A. J. Matheson (1967). *J. Chem. Phys.*, **46**, 2451.

Cauchy, A. L. (1882). Quoted by Zwanzig and Mountain (1965).

Cerceo, M., R. Meister and T. A. Litovitz (1962). *J. Acoust. Soc. Am.*, **34**, 259.

Cerf, R. (1952). *Compt. Rend.*, **234**, 1549.

Cerf, R. and G. B. Thurston (1964). *J. Chim. Phys.*, **61**, 1457.
Chan, S. C. and J. P. Valleau (1968). *Canad. J. Chem.*, **46**, 853.
Chase, C. E. (1958). *Phys. Fluids*, **1**, 193.
Chase, C. E., R. C. Williamson and L. Tisza (1964). *Phys. Rev. Lett.*, **13**, 467.
Chen, J. H. and A. A. Petrauskas (1959). *J. Chem. Phys.*, **30**, 304.
Chiao, R. Y. and B. P. Stoicheff (1964). *J. Opt. Soc. Am.*, **54**, 1286.
Chiao, R. Y., C. H. Townes and B. P. Stoicheff (1964). *Phys. Rev. Lett.*, **12**, 592.
Chmelnick, A. M. and D. Fiat (1967). *J. Chem. Phys.*, **47**, 3986.
Chompff, A. J. and J. A. Duiser (1966). *J. Chem. Phys.*, **45**, 1505.
Chompff, A. J. and W. Prins (1968). *J. Chem. Phys.*, **48**, 235.
Chynoweth, A. G. and W. G. Schneider (1952). *J. Chem. Phys.*, **20**, 1777.
Clark, A. E. and T. A. Litovitz (1960). *J. Acoust. Soc. Am.*, **32**, 1221.
Clark, N. A., C. E. Moeller, J. A. Bucaro and E. F. Carome (1966). *J. Chem. Phys.*, **44**, 2528.
Cohen, M. H. and D. Turnbull (1959). *J. Chem. Phys.*, **31**, 1164.
Cole, K. S. and R. H. Cole (1941). *J. Chem. Phys.*, **9**, 341.
Comley, W. J. (1966). *Brit. J. Appl. Phys.*, **17**, 1375.
Corran, P. G., J. D. Lambert, R. Salter and B. Warburton (1958). *Proc. Roy. Soc. A*, **244**, 212.
Cottrell, A. H. (1953). *Dislocations and Plastic Flow in Crystals*, Oxford University Press, Oxford.
Cottrell, T. L. and N. Ream (1955). *Trans. Faraday Soc.*, **51**, 159, 1453.
Cottrell, T. L. and J. C. McCoubrey (1961). *Molecular Energy Transfer in Gases*, Butterworths, London.
Cottrell, T. L. and A. J. Matheson (1962). *Trans. Faraday Soc.*, **58**, 2336.
Cottrell, T. L. and A. J. Matheson (1963). *Trans. Faraday Soc.*, **59**, 824.
Cottrell, T. L., R. C. Dobbie, J. McLain and A. W. Read (1964). *Trans. Faraday Soc.*, **60**, 241.
Cottrell, T. L. (1965). *Dynamic Aspects of Molecular Energy States*, Oliver and Boyd, Edinburgh.
Cottrell, T. L., I. M. Macfarlane and A. W. Read (1965). *Trans. Faraday Soc.*, **61**, 1632.
Cottrell, T. L. and M. A. Day (1966). *Chem. Soc. Spec. Publ.*, **20**, 253.
Cottrell, T. L., I. M. Macfarlane, A. W. Read and A. H. Young (1966). *Trans Faraday Soc.*, **62**, 2655.
Cox, W. P., L. E. Nielsen and R. Keeney (1957). *J. Polymer Sci.*, **26**, 365.
Craft, W. L. and L. J. Slutsky (1968). *J. Chem. Phys.*, **49**, 638.
Crapo, L. M. and G. W. Flynn (1965). *J. Chem. Phys.*, **43**, 1443.
Crook, K. R. and E. Wyn-Jones (1969). *J. Chem. Phys.*, **50**, 3445.
Curie, P. and J. Curie (1880). *Compt. Rend.*, **91**, 294.
Curl, R. F. (1959). *J. Chem. Phys.*, **30**, 1529.
Curtis, B. J. (1969). *J. Chem. Phys.*, **40**, 433.
Czerlinski, G. H. (1966). *Chemical Relaxation*, Arnold, London.
Darbari, G. S. and S. Petrucci (1970). *J. Phys. Chem.*, **74**, 268.
D'Arrigo, G. and D. Sette (1968). *J. Chem. Phys.*, **48**, 691.
Davidson, D. W. and R. H. Cole (1951), *J. Chem. Phys.*, **19**, 1484.
Davies, D. B. and A. J. Matheson (1966). *J. Chem. Phys.*, **45**, 1000.
Davies, D. B. and A. J. Matheson (1967). *Trans. Faraday Soc.*, **63**, 596.
Davies, D. B. (1968). Thesis, University of Essex.
Davies, R. O. and J. Lamb (1959). *Proc. Phys. Soc.*, **73**, 767.

Davison, W. D. (1962). *Disc. Faraday Soc.*, **33**, 71.

Debye, P. (1912). *Ann. Physik*, **39**, 789.

Debye, P. and F. W. Sears (1932). *Proc. Nat. Acad. Sci. U.S.*, **18**, 409.

Debye, P. (1933). *J. Chem. Phys.*, **1**, 13.

De Mallie, R. B., M. H. Birnboim, J. E. Frederick, N. W. Tschoegl and J. D. Ferry (1962). *J. Phys. Chem.*, **66**, 536.

DeMartini, F. and J. Ducuing (1966). *Phys. Rev. Lett.*, **17**, 117.

Dev, S. B., R. Y. Lochhead and A. M. North (1970). *Disc. Faraday Soc.*, **49**, in the press.

Dewey, J. M. (1947). *J. Appl. Phys.*, **18**, 578.

Dexter, A. R. and A. J. Matheson (1968). *Trans. Faraday Soc.*, **64**, 2632.

Dexter, A. R. (1970). Thesis, University of Essex.

Dexter, A. R. and A. J. Matheson (1970). *J. Chem. Phys.*, **53**, in the press.

Dickie, R. A. and J. D. Ferry (1966). *J. Phys. Chem.*, **70**, 2594.

Dienes, G. J. (1958). *J. Phys. Chem. Solids*, **7**, 290.

Diller, D. E. (1965). *J. Chem. Phys.*, **42**, 2089.

Doolittle, A. K. (1951). *J. Appl. Phys.*, **22**, 1471.

Doolittle, A. K. and D. B. Doolittle (1957). *J. Appl. Phys.*, **28**, 901.

Doyennette, L. and L. Henry (1966). *J. Phys. Radium*, **27**, 485.

Duff, A. W. (1898). *Phys. Rev.*, **6**, 129.

Dunn, F. and J. E. Breyer (1962). *J. Acoust. Soc. Am.*, **34**, 775.

Eby, R. K. (1964). *J. Acoust. Soc. Am.*, **36**, 1485.

Eccleston, G., R. A. Pethrick, E. Wyn-Jones, P. C. Hamblin and R. F. M. White (1970). *Trans. Faraday Soc.*, **66**, 310.

Edmonds, P. D. and J. Lamb (1958). *Proc. Phys. Soc.*, **72**, 940.

Edmonds, P. D., V. F. Pearce and J. H. Andreae (1962). *Brit. J. Appl. Phys.*, **13**, 551.

Edmonds, P. D. (1966). *Rev. Sci. Inst.*, **37**, 367.

Edwards, C., J. A. Lipa and M. J. Buckingham (1968). *Phys. Rev. Lett.*, **20**, 496.

Eggers, F. (1968). *Acustica*, **19**, 323.

Eigen, M., K. Kustin and G. Maass (1961). *Z. Phys. Chem.*, **30**, 130.

Eigen, M. and K. Tamm (1962). *Z. Elektrochem.*, **66**, 93.

Einstein, A. (1906). *Ann. Physik*, **19**, 289.

Einstein, A. (1911). *Ann. Physik*, **34**, 591.

Einstein, A. (1920). *Preuss. Akad. Wiss., Berlin, Ber.*, **24**, 380.

Eliel, E. L. (1962). *Stereochemistry of Carbon Compounds*, McGraw–Hill, New York.

Ener, C., A. Busala and J. C. Hubbard (1955). *J. Chem. Phys.*, **23**, 155.

Erginsav, A. (1969). Thesis, University of Glasgow.

Eucken, A. and H. Jaacks (1935). *Z. Phys. Chem. B*, **30**, 85.

Eucken, A. and E. Nümann (1937). *Z. Phys. Chem. B*, **36**, 163.

Eucken, A. and S. Aybar (1940). *Z. Phys. Chem. B*, **46**, 195.

Eyring, H. (1936). *J. Chem. Phys.*, **4**, 283.

Eyring, H. and J. Hirschfelder (1937). *J. Phys. Chem.*, **41**, 249.

Fabelinskii, I. L. (1957). *Uspekhi Fiz. Nauk*, **63**, 355.

Fay, D. P., D. Litchinsky and N. Purdie (1969). *J. Phys. Chem.*, **73**, 544.

Fay, D. P. and N. Purdie (1970). *J. Phys. Chem.*, **74**, 1160.

Ferguson, M. G. and A. W. Read (1965). *Trans. Faraday Soc.*, **61**, 1559.

Ferguson, M. G. and A. W. Read (1967). *Trans. Faraday Soc.*, **63**, 61.

Ferry, J. D. (1941). *Rev. Sci. Inst.*, **12**, 79.

Ferry, J. D., M. L. Williams and D. M. Stern (1954). *J. Phys. Chem.*, **58**, 987.

Ferry, J. D., R. F. Landel and M. L. Williams (1955). *J. Appl. Phys.*, **26**, 359.

Ferry, J. D. (1961). *Viscoelastic Properties of Polymers*, Wiley, New York.

Ferry, J. D., L. A. Holmes, J. Lamb and A. J. Matheson (1966). *J. Phys. Chem.*, **70**, 1685.

Ferry, J. D. (1969). *Proc. 5th Int. Cong. Rheology*, **1**, 3.

Fitzgerald, E. R., L. D. Grandine and J. D. Ferry (1953). *J. Appl. Phys.*, **24**, 650.

Fixman, M. (1961). *J. Chem. Phys.*, **34**, 369.

Fixman, M. (1962). *J. Chem. Phys.*, **36**, 1961.

Fleury, P. A. and R. Y. Chiao (1966). *J. Acoust. Soc. Am.*, **39**, 751.

Fogg, P. G. T., P. A. Hanks and J. D. Lambert (1953). *Proc. Roy. Soc. A*, **219**, 490.

Foster, N. F. (1969). *J. Appl. Phys.*, **40**, 420.

Fox, T. G., S. Gratch and S. Loshaek (1956). In F. R. Eirich (Ed.), *Rheology*, Vol. 1, Academic Press, New York.

Frederick, J. E., N. W. Tschoegl and J. D. Ferry (1964). *J. Phys. Chem.*, **68**, 1974.

Frederick, J. E. and J. D. Ferry (1965). *J. Phys. Chem.*, **69**, 346.

Fritsch, K., C. J. Montrose, J. L. Hunter and J. F. Dill (1970). *J. Chem. Phys.*, **52**, 2242.

Fröhlich, H. (1949). *Theory of Dielectrics*, Oxford University Press, Oxford.

Fujii, Y., R. B. Lindsay and K. Urushihara (1963). *J. Acoust. Soc. Am.*, **35**, 961.

Fulcher G. S. (1925). *J. Am. Ceram. Soc.*, **8**, 339.

Fuoss, R. M. and J. G. Kirkwood (1941). *J. Am. Chem. Soc.*, **63**, 385.

Gabrielli, I. and L. Verdini (1955). *Nuov. Cim.*, **2**, 526.

Garland, C. W. and C. F. Yarnell (1966). *J. Chem. Phys.*, **44**, 3678.

Garza, V. L. and N. Purdie (1970). *J. Phys. Chem.*, **74**, 275.

Gauster, W. B. and M. A. Breazale (1968). *Phys. Rev.*, **168**, 655.

Generalov, N. A. (1963). *Doklady Akad. Nauk S.S.S.R.*, **148**, 373.

Gitis, M. B. and A. S. Khimunin (1969). *Sov. Phys. Acoust.*, **14**, 413. [(1968). *Akust. Zh.*, **14**, 489.]

Glarum, S. H. (1960). *J. Chem. Phys.*, **33**, 639.

Glasstone, S., K. J. Laidler and H. Eyring (1941). *Theory of Rate Processes*, McGraw–Hill, New York.

Glover, G. M., G. Hall, A. J. Matheson and J. L. Stretton (1968). *J. Phys. E*, **1**, 383.

Glover, G. M., G. Hall, A. J. Matheson and J. L. Stretton (1969). *Proc. 5th Int. Cong. Rheology*, **1**, 429.

Goldblatt, N. R. and T. A. Litovitz (1967). *J. Acoust. Soc. Am.*, **41**, 1301.

Golding, B. and M. Barmatz (1969). *Phys. Rev. Lett.*, **23**, 223.

Goldstein, M. (1969). *J. Chem. Phys.*, **51**, 3728.

Gooberman, G. L. (1968). *Ultrasonics*, English Universities Press, London.

Gordon, R. G., W. Klemperer and J. I. Steinfeld (1968). *Ann. Rev. Phys. Chem.*, **19**, 215.

Gorelik, G. (1946). *Doklady Akad. Nauk S.S.S.R.*, **54**, 779.

Granato, A. and K. Lücke (1956). *J. Appl. Phys.*, **27**, 583.

Grandine, L. D. and J. D. Ferry (1953). *J. Appl. Phys.*, **24**, 679.

Gravitt, J. C. (1960). *J. Acoust. Soc. Am.*, **32**, 560.

Gray, P. and S. A. Rice (1964). *J. Chem. Phys.*, **41**, 3689.

Green, H. S. (1952). *The Molecular Theory of Fluids*, North Holland, Amsterdam.

Greenspan, M. (1950). *J. Acoust. Soc. Am.*, **22**, 568.

Greenspan, M. and M. C. Thompson (1953). *J. Acoust. Soc. Am.*, **25**, 92.

Greenspan, M. (1954). *J. Acoust. Soc. Am.*, **26**, 70.

Greenspan, M. (1956). *J. Acoust. Soc. Am.*, **28**, 644.

Greenspan, M. and C. E. Tschiegg (1957). *Rev. Sci. Inst.*, **28**, 897.

Greenspan, M. (1965). In W. P. Mason (Ed.), *Physical Acoustics*, Vol. 2A, Academic Press, New York.

de Groot, M. S. and J. Lamb (1957). *Proc. Roy. Soc. A*, **242**, 36.

Gross, B. (1968). *Mathematical Structure of the Theories of Viscoelasticity*, Hermann, Paris.

Gruber, G. J. and T. A. Litovitz (1964). *J. Chem. Phys.*, **40**, 13.

Guth, E. and H. Mark (1934). *Monatsh. Chem.*, **65**, 93.

Haebel, E. U. (1965). *Acustica*, **15**, 426.

Haebel, E. U. (1968). *Acustica*, **20**, 65.

Hakim, S. E. A. and W. J. Comley (1965). *Nature*, **208**, 1082.

Hall, D. N. (1959). *Trans. Faraday Soc.*, **55**, 1319.

Hall, D. N. and J. Lamb (1959). *Proc. Phys. Soc.*, **73**, 354.

Hamblin, P. C., R. F. M. White and E. Wyn-Jones (1968). *Chem. Comm.*, 1058.

Hamblin, P. C., R. F. M. White and E. Wyn-Jones (1969). *J. Molec. Struct.*, **4**, 275.

Hammerle, W. G. and J. G. Kirkwood (1955). *J. Chem. Phys.*, **23**, 1743.

Hammes, G. G. and H. O. Spivey (1966). *J. Am. Chem. Soc.*, **88**, 1621.

Hammes, G. G. and C. N. Pace (1968). *J. Phys. Chem.*, **72**, 2227.

Hammes, G. G. and A. C. Park (1968). *J. Am. Chem. Soc.*, **90**, 4151.

Hancock, J. K. and J. C. Decius (1969). *J. Chem. Phys.*, **51**, 5374.

Harlow, R. G. and M. E. Nolan (1967). *J. Acoust. Soc. Am.*, **42**, 899.

Harris, R. A. and J. E. Hearst (1966). *J. Chem. Phys.*, **44**, 2595.

Hässler, H. and H. J. Bauer (1969). *Kolloid Z.*, **230**, 194.

Hawley, S. A. and F. Dunn (1969). *J. Chem. Phys.*, **50**, 3523.

Hearst, J. E. (1962). *J. Chem. Phys.*, **37**, 2547.

Heasell, E. L. and J. Lamb (1956a). *Proc. Roy. Soc. A*, **237**, 233.

Heasell, E. L. and J. Lamb (1956b). *Proc. Phys. Soc. B*, **69**, 869.

Heatley, F. and G. Allen (1969). *Molec. Phys.*, **16**, 77.

Heijboer, J. (1965). In J. A. Prins (Ed.), *Physics of Non-crystalline Solids*, North Holland, Amsterdam.

Heijboer, J. (1968). *J. Polymer Sci. C*, **16**, 3413.

Hemphill, R. B. (1968). *Rev. Sci. Inst.*, **39**, 910.

Henderson, M. C. and L. Peselnick (1957). *J. Acoust. Soc. Am.*, **29**, 1074.

Henderson, M. C. (1962). *J. Acoust. Soc. Am.*, **34**, 349.

Henderson, M. C. and G. J. Donnelly (1962). *J. Acoust. Soc. Am.*, **34**, 779.

Henderson, M. C., K. F. Herzfeld, J. Bry, R. Coakley and G. Carriere (1969a). *J. Acoust. Soc. Am.*, **45**, 109.

Henderson, M. C., L. J. Burbank and J. J. Glatzel (1969b). *J. Acoust. Soc. Am.*, **46**, 819.

Herzberg, G. (1945). *Molecular Spectra and Molecular Structure*, Vol. 2, Van Nostrand, New York.

Herzfeld, K. F. and F. O. Rice (1928). *Phys. Rev.*, **31**, 691.

Herzfeld, K. F. (1952). *J. Chem. Phys.*, **20**, 288.

Herzfeld, K. F. (1957). *J. Acoust. Soc. Am.*, **29**, 1180.

Herzfeld, K. F. and T. A. Litovitz (1959). *Absorption and Dispersion of Ultrasonic Waves*, Academic Press, New York.

Herzfeld, K. F. (1962). *J. Chem. Phys.*, **36**, 3305.

Herzfeld, K. F. (1966). *J. Acoust. Soc. Am.*, **39**, 813.

Higgs, R. W. and T. A. Litovitz (1960). *J. Acoust. Soc. Am.*, **32**, 1108.

Hill, G. L. and T. G. Winter (1968). *J. Chem. Phys.*, **49**, 440.

Hinsch, H. (1961). *Acustica*, **11**, 230, 426.

Hirai, N. and H. Eyring (1958). *J. Appl. Phys.*, **29**, 810.

Holmes, L. A. and J. D. Ferry (1968). *J. Polymer Sci. C*, **23**, 291.

Holmes, R., F. A. Smith and W. Tempest (1963). *Proc. Phys. Soc.*, **81**, 311.

Holmes, R., G. R. Jones and R. Lawrence (1964). *J. Chem. Phys.*, **41**, 2955.

Holmes, R., G. R. Jones and N. Pusat (1964a). *Trans. Faraday Soc.*, **60**, 1220.

Holmes, R., G. R. Jones and N. Puset (1964b). *J. Chem. Phys.*, **41**, 2512.

Holmes, R., G. R. Jones and R. Lawrence (1966). *Trans. Faraday Soc.*, **62**, 46.

Hooker, W. J. and R. C. Millikan (1963). *J. Chem. Phys.*, **38**, 214.

Hubbard, J. C. and A. L. Loomis (1927). *Nature*, **120**, 189.

Hubbard, J. C. (1931). *Phys. Rev.*, **38**, 1011.

Huber, P. W. and A. Kantrowitz (1947). *J. Chem. Phys.*, **15**, 275.

Hudson, G. H., J. C. McCoubrey and A. R. Ubbelohde (1961). *Proc. Roy. Soc. A*, **264**, 289.

Hunt, B. I. and J. G. Powles (1966). *Proc. Phys. Soc.*, **88**, 513.

Hunter, J. L. and H. D. Dardy (1965). *J. Chem. Phys.*, **42**, 2961.

Hunter, J. L. and H. D. Dardy (1966). *J. Chem. Phys.*, **44**, 3637.

Hunter, J. L. and P. R. Derdul (1967). *J. Acoust. Soc. Am.*, **42**, 1041.

Hutson, A. R., J. H. McFee and D. L. White (1961). *Phys. Rev. Lett.*, **7**, 237.

Hutton, J. F. (1968). *Proc. Roy. Soc. A*, **304**, 65.

Hutton, J. F. and M. C. Phillips (1969). *J. Chem. Phys.*, **51**, 1065.

Illers, K. H. and E. Jenckel (1959). *J. Polymer Sci.*, **41**, 528.

Illers, K. H. (1961). *Z. Elektrochem.*, **65**, 679.

Imai, J. S. and I. Rudnick (1969). *J. Acoust. Soc. Am.*, **46**, 1144.

Inamura, T. (1970). *Japan. J. Appl. Phys.*, **9**, 255.

Isakovich, M. A. and I. A. Chaban (1966a). *Sov. Phys. Doklady*, **10**, 1055. [(1965). *Doklady Akad. Nauk S.S.S.R.*, **165**, 299.]

Isakovich, M. A. and I. A. Chaban (1966b). *Sov. Phys. J.E.T.P.*, **23**, 893. [*Zh. Eksp. Teor. Fiz.*, **50**, 1343.]

Jackson, J. M. and N. F. Mott (1932). *Proc. Roy. Soc. A*, **137**, 703.

Jarzynski, J. (1963). *Proc. Phys. Soc.*, **81**, 745.

Jarzynski, J. and T. A. Litovitz (1964). *J. Chem. Phys.*, **41**, 1290.

Jones, D. G., J. D. Lambert, M. P. Saksena and J. L. Stretton (1969). *Trans. Faraday Soc.*, **65**, 965.

Jonkman, R. M., G. J. Prangsma, I. Ertas, H. F. P. Knaap and J. J. M. Beenakker (1968a). *Physica*, **38**, 441.

Jonkman, R. M., G. J. Prangsma, R. A. J. Keijser, R. A. Aziz and J. J. M. Beenakker (1968b). *Physica*, **38**, 451.

Jonkman, R. M., G. J. Prangsma, R. A. J. Keijser, H. F. P. Knaap and J. J. M. Beenakker (1968c). *Physica*, **38**, 456.

Karpovich, J. (1954). *J. Chem. Phys.*, **22**, 1767.

Kästner, S. (1962a). *Kolloid Z.*, **178**, 119.

Kästner, S. (1962b). *Kolloid Z.*, **184**, 109.

Kaulgud, M. V. (1965). *Acustica*, **15**, 377.

Kauzmann, W. (1948). *Chem. Rev.*, **43**, 219.

Kelley, J. D. and M. Wolfsberg (1966a). *J. Chem. Phys.*, **44**, 324.

Kelley, J. D. and M. Wolfsberg (1966b). *J. Chem. Phys.*, **45**, 3881.
Kelpin, H. and O. Weis (1967). *Acustica*, **18**, 105.
Kessler, L. W. and F. Dunn (1969). *J. Phys. Chem.*, **73**, 4256.
Khabibullaev, P. K. and M. G. Khaliulin (1969). *Sov. Phys. Acoust.*, **15**, 120. [*Akust. Zh.*, **15**, 140.]
Kiefer, J. H. and R. W. Lutz (1966). *J. Chem. Phys.*, **45**, 3888.
Kiefer, J. H., W. D. Breshears and P. F. Bird (1969). *J. Chem. Phys.*, **50**, 3641.
Kinsler, L. E. and A. R. Frey (1962). *Fundamentals of Acoustics*, Wiley, New York.
Kirchhoff, G. (1868). *Pogg. Ann. Phys.*, **134**, 177.
Kirkwood, J. G. (1946). *J. Chem. Phys.*, **14**, 51.
Kirkwood, J. G. and J. Riseman (1948). *J. Chem. Phys.*, **16**, 565.
Kirkwood, J. G. and P. L. Auer (1951). *J. Chem. Phys.*, **19**, 281.
Kittel, C. (1966). *Introduction to Solid State Physics*, Wiley, New York.
deKlerk, J. and E. F. Kelly (1965). *Rev. Sci. Inst.*, **36**, 506.
Kneser, H. O. (1938). *Ann. Physik*, **32**, 277.
Knollman, G. C., D. O. Miles and A. S. Hamamoto (1965). *J. Chem. Phys.*, **43**, 1160.
Kohler, M. (1941). *Ann. Physik*, **39**, 209.
Kohler, M. (1950). *Z. Phys.*, **127**, 41.
Kono, R., G. E. McDuffie and T. A. Litovitz (1966). *J. Chem. Phys.*, **44**, 965.
Kono, R., T. A. Litovitz and G. E. McDuffie (1966). *J. Chem. Phys.*, **45**, 1790.
Kor, S. K., G. Rai and O. N. Awasthi (1969). *Phys. Lett. A*, **30**, 289.
Kovacs, A. J. (1958). *J. Polymer Sci.*, **30**, 131.
Krasnooshkin, P. E. (1944). *Phys. Rev.*, **65**, 190.
Kratky, O. and G. Porod (1949). *Rec. Trav. Chim.*, **68**, 1106.
Krebs, K. and J. Lamb (1958). *Proc. Roy. Soc. A*, **244**, 558.
Kuhl, W., G. R. Schodder and F. K. Schröder (1954). *Acustica*, **4**, 519.
Kuhn, W. (1934). *Kolloid Z.*, **68**, 2.
Kumar, S. F. (1963). *Phys. Chem. Glasses*, **4**, 106.
Kurtze, G. and K. Tamm (1953). *Acustica*, **3**, 33.
Lamb, J. and J. M. M. Pinkerton (1949). *Proc. Roy. Soc. A*, **199**, 114.
Lamb, J. and J. H. Andreae (1951). *Nature*, **167**, 898.
Lamb, J. and J. Sherwood (1955). *Trans. Faraday Soc.*, **51**, 1674.
Lamb, J. (1960). *Z. Elektrochem.*, **64**, 135.
Lamb, J. and A. J. Matheson (1964). *Proc. Roy. Soc. A*, **281**, 207.
Lamb, J. (1965). In W. P. Mason (Ed.), *Physical Acoustics*, Vol. 2A, Academic Press, New York.
Lamb, J. and J. Richter (1966). *Proc. Roy. Soc. A*, **293**, 479.
Lamb, J. and H. Seguin (1966). *J. Acoust. Soc. Am.*, **39**, 519.
Lamb, J. and P. Lindon (1967). *J. Acoust. Soc. Am.*, **41**, 1032.
Lambert, J. D. and J. S. Rowlinson (1951). *Proc. Roy. Soc. A*, **204**, 424.
Lambert, J. D. and R. Salter (1957). *Proc. Roy. Soc. A*, **243**, 78.
Lambert, J. D. and R. Salter (1959). *Proc. Roy. Soc. A*, **253**, 277.
Lambert, J. D., A. J. Edwards, D. Pemberton and J. L. Stretton (1962). *Disc. Faraday Soc.*, **33**, 61.
Lambert, J. D., D. G. Parks-Smith and J. L. Stretton (1964). *Proc. Roy. Soc. A*, **282**, 380.
Lambert, J. D. (1967). *Quart. Rev. Chem. Soc.*, **21**, 67.
Landau, L. and E. Teller (1936). *Phys. Z. Soviet*, **10**, 34.

Landel, R. F. and J. D. Ferry (1955). *J. Phys. Chem.*, **59**, 658.

Landolt-Börnstein (1967). *Zahlenwerte und Funktionen aus Naturwissenschaften und Technik*, Vol. II/5, Ed. W. Schaaffs, Springer, Berlin.

Law, A. K., N. Koronaios and R. B. Lindsay (1967). *J. Acoust. Soc. Am.*, **41**, 93.

LeCraw, R. C. and R. L. Comstock (1965). In W. P. Mason (Ed.), *Physical Acoustics*, Vol. 3B, Academic Press, New York.

Leidecker, H. W. and J. T. LaMacchia (1968). *J. Acoust. Soc. Am.*, **43**, 143.

Lestz, S. S. and R. N. Grove (1965). *J. Chem. Phys.*, **43**, 883.

Levy, M. and I. Rudnick (1962). *J. Acoust. Soc. Am.*, **34**, 520.

Lide, D. R. (1964). *Ann. Rev. Phys. Chem.*, **15**, 225.

Liebermann, L. (1959). *Phys. Rev.*, **113**, 1052.

Litovitz, T. A., T. Lyon and L. Peselnick (1954). *J. Acoust. Soc. Am.*, **26**, 566.

Litovitz, T. A. and T. Lyon (1954). *J. Acoust. Soc. Am.*, **26**, 577.

Litovitz, T. A., E. H. Carnevale and P. A. Kendall (1957). *J. Chem. Phys.*, **26**, 465.

Litovitz, T. A. (1957). *J. Chem. Phys.*, **26**, 469.

Litovitz, T. A. and E. H. Carnevale (1958). *J. Acoust. Soc. Am.*, **30**, 134.

Litovitz, T. A. and C. M. Davis (1965). In W. P. Mason (Ed.), *Physical Acoustics*, Vol. 2A, Academic Press, New York.

Llewellyn, J. D., H. M. Montagu-Pollock and E. R. Dobbs (1969). *J. Phys. E*, **2**, 535.

Longuet-Higgins, H. C. and J. P. Valleau (1958). *Molec. Phys.*, **1**, 284.

Lorentz, H. A. (1881). *Arc. Neerl.*, **16**, 1. Reprinted in P. Zeeman and A. D. Fokker (Eds.) (1938), *Collected Papers of H. A. Lorentz*, Vol. 6, Martinus Nijhoff, The Hague.

Lovell, S. E. and J. D. Ferry (1961). *J. Phys. Chem.*, **65**, 2274.

Lowe, J. P. (1968). *Prog. Phys. Org. Chem.*, **6**, 1.

Lucas, R. and P. Biquard (1932). *J. Phys. Radium*, **3**, 464.

Lucretius Carus, T. (99–55 B.C.). *De Rerum Natura*.

Lyon, T. and T. A. Litovitz (1956). *J. Appl. Phys.*, **27**, 179.

McCoubrey, J. C., J. B. Parke and A. R. Ubbelohde (1954). *Proc. Roy. Soc. A*, **223**, 155.

McCoubrey, J. C., R. C. Milward and A. R. Ubbelohde (1961a). *Proc. Roy. Soc. A*, **264**, 299.

McCoubrey, J. C., R. C. Milward and A. R. Ubbelohde (1961b). *Trans. Faraday Soc.*, **57**, 1472.

McCoubrey, J. C., R. C. Milward and A. R. Ubbelohde (1962). *Proc. Roy. Soc. A*, **269**, 456.

McCrum, N. G., B. E. Read and G. Williams (1967). *Anelastic and Dielectric Effects in Polymeric Solids*, Wiley, London.

McDuffie, G. E. and T. A. Litovitz (1962). *J. Chem. Phys.*, **37**, 1699.

Macedo, P. and T. A. Litovitz (1965). *Phys. Chem. Glasses*, **6**, 69.

McFee, J. H. (1966). In W. P. Mason (Ed.), *Physical Acoustics*, Vol. 4A, Academic Press, New York.

McKellar, J. F. and J. H. Andreae (1962). *Nature*, **195**, 778.

McSkimin, H. J. (1952). *J. Acoust. Soc. Am.*, **24**, 355.

McSkimin, H. J. (1957). *J. Acoust. Soc. Am.*, **29**, 1185.

McSkimin, H. J. (1960). *J. Acoust. Soc. Am.*, **32**, 1401.

McSkimin, H. J. and P. Andreatch (1967). *J. Acoust. Soc. Am.*, **42**, 248.

McSkimin, H. J. and T. B. Bateman (1969). *J. Acoust. Soc. Am.*, **45**, 852.

Madigosky, W. M. and T. A. Litovitz (1961). *J. Chem. Phys.*, **34**, 489.

Madigosky, W. M. (1963). *J. Chem. Phys.*, **39**, 2704.

Madigosky, W. M., G. E. McDuffie and T. A. Litovitz (1967). *J. Chem. Phys.*, **47**, 753.

Madigosky, W. M., A. Monkewicz and T. A. Litovitz (1967). *J. Acoust. Soc. Am.*, **41**, 1308.

Maekawa, E., R. G. Mancke and J. D. Ferry (1965). *J. Phys. Chem.*, **69**, 2811.

Mancke, R. G., R. A. Dickie and J. D. Ferry (1968). *J. Polymer Sci. A2*, **6**, 1783.

Markovitz, H., T. G. Fox and J. D. Ferry (1962). *J. Phys. Chem.*, **66**, 1567.

Marriott, R. (1964). *Proc. Phys. Soc.*, **83**, 159.

Martinez, J. V., J. G. Strauch and J. C. Decius (1964). *J. Chem. Phys.*, **40**, 186.

Mason, E. A. and L. Monchick (1962). *J. Chem. Phys.*, **36**, 1622.

Mason, W. P. (1947). *Trans. Am. Soc. Mech. Eng.*, **69**, 359.

Mason, W. P. and H. J. McSkimin (1947). *J. Acoust. Soc. Am.*, **19**, 464.

Mason, W. P., W. O. Baker, H. J. McSkimin and J. H. Heiss (1948). *Phys. Rev.*, **73**, 1074.

Mason, W. P. and H. J. McSkimin (1948). *J. Appl. Phys.*, **19**, 940.

Mason, W. P., W. O. Baker, H. J. McSkimin and J. H. Heiss (1949). *Phys. Rev.*, **75**, 936.

Mason, W. P. (1950). *Piezoelectric Crystals and their Application to Ultrasonics*, Van Nostrand, New York.

Mason, W. P. (1958). *Physical Acoustics and the Properties of Solids*, Van Nostrand, New York.

Mason, W. P. and T. B. Bateman (1963). *Phys. Rev. Lett.*, **10**, 151.

Mason, W. P. (Ed.) (1964–1970). *Physical Acoustics*, Vols. 1–6, Academic Press, New York.

Matheson, A. J. (1962). *Disc. Faraday Soc.*, **33**, 94.

Maxwell, J. C. (1867). *Phil. Trans. Roy. Soc.*, **157**, 52.

Meister, R., C. J. Marhoeffer, R. Sciamanda, L. Cotter and T. A. Litovitz (1960). *J. Appl. Phys.*, **31**, 854.

Melcher, R. L., D. I. Bolef and J. B. Merry (1968). *Rev. Sci. Inst.*, **39**, 1613.

Meyer, E. and G. Sessler (1957). *Z. Phys.*, **149**, 15.

Meyer, H. H., W. F. Pfeiffer and J. D. Ferry (1967). *Biopolymers*, **5**, 123.

Meyer, N. J. (1960). *J. Chem. Phys.*, **33**, 487.

Mies, F. H. (1965). *J. Chem. Phys.*, **42**, 2709.

Mikhailov, I. G. and L. I. Tarutina (1950). *Doklady Akad. Nauk S.S.S.R.*, **74**, 41.

Mikhailov, I. G. (1963). *Sov. Phys. Acoust.*, **8**, 375. [(1962). *Akust. Zh.*, **8**, 478.]

Mikhailov, I. G. (1964). *Ultrasonics*, **2**, 203.

Millikan, R. C. and D. R. White (1963). *J. Chem. Phys.*, **39**, 98.

Millikan, R. C. (1964). *J. Chem. Phys.*, **40**, 2594.

Millikan, R. C. and L. A. Osburg (1964). *J. Chem. Phys.*, **41**, 2196.

Milward, R. C. and A. R. Ubbelohde (1963). *Proc. Roy. Soc. A*, **272**, 481.

Mizushima, S. (1954). *The Structure of Molecules and Internal Rotation*, Academic Press, New York.

Montrose, C. J., V. A. Solovyev and T. A. Litovitz (1968). *J. Acoust. Soc. Am.*, **43**, 117.

Montrose, C. J. and K. Fritsch (1970). *J. Acoust. Soc. Am.*, **47**, 786.

Montrose, C. J. and T. A. Litovitz (1970). *J. Acoust. Soc. Am.*, **47**, 1250.

Moore, C. B. (1965). *J. Chem. Phys.* **43**, 2979.

Moore, R. S., H. J. McSkimin, C. Gieniewski and P. Andreatch (1967a). *J. Chem. Phys.* **47**, 3.

Moore, R. S., H. J. McSkimin, C. Gieniewski and P. Andreatch (1967b). *J. Chem. Phys.*, **47**, 4329.

Moore, R. S., H. J. McSkimin, C. Gieniewski and P. Andreatch (1969). *J. Chem. Phys.*, **50**, 466.

Morita, E., R. Kono and H. Yoshizaki (1968). *Japan. J. Appl. Phys.*, **7**, 451.

Morrisson, T. E., L. J. Zapas and T. W. DeWitt (1955). *Rev. Sci. Inst.*, **26**, 357.

Mountain, R. D. (1966). *J. Chem. Phys.*, **44**, 832.

Mountain, R. D. and R. Zwanzig (1966). *J. Chem. Phys.*, **44**, 2777.

Müller, F. H. (1969). *Proc. 5th Int. Cong. Rheology*, **1**, 61.

Naugle, D. G., J. H. Lunsford and J. R. Singer (1966). *J. Chem. Phys.*, **45**, 4669.

Neklepajew, N. (1911). *Ann. Physik*, **35**, 175.

Nichols, W. H. and E. F. Carome (1968). *J. Chem. Phys.*, **49**, 1000.

Nine, H. D. and R. Truell (1961). *Phys. Rev.* **123**, 799.

Nomura, H., S. Kato and Y. Miyahara (1969). *J. Chem. Soc. Japan*, **90**, 250.

Nozdrev, V. F. (1965). *Use of Ultrasonics in Molecular Physics*, Pergamon, Oxford.

Nyeland, C. (1967). *J. Chem. Phys.*, **46**, 63.

Ohsawa, T. and Y. Wada (1967). *Japan. J. Appl. Phys.*, **6**, 1351.

Olson, J. R. and S. Legvold (1963). *J. Chem. Phys.*, **39**, 2902.

O'Neal, C. and R. S. Brokaw (1963). *Phys. Fluids*, **6**, 1675.

O'Neil, H. T. (1949). *Phys. Rev.*, **75**, 928.

Orville-Thomas, W. J. and E. Wyn-Jones (1969). In G. M. Burnett and A. M. North (Eds.), *Transfer and Storage of Energy by Molecules*, Vol. 2, Wiley, London, p. 265.

Owen, N. L. and N. Sheppard (1963). *Proc. Chem. Soc.*, 264.

Padmanaban, R. A. (1960). *J. Sci. Ind. Res. India B*, **19**, 336.

Pancholy, M. and S. S. Mathur (1963). *Acustica*, **13**, 42.

Pancholy, M. and T. K. Saksena (1967). *Acustica*, **18**, 299.

Pancholy, M. and T. K. Saksena (1967). *J. Phys. Soc. Japan*, **22**, 1110.

Parbrook, H. D. and E. G. Richardson (1952). *Proc. Phys. Soc. B*, **65**, 437.

Park, P. J. D. and E. Wyn-Jones (1968). *J. Chem. Soc. A*, 2064.

Parker, J. G., C. E. Adams and R. M. Stavseth (1953). *J. Acoust. Soc. Am.*, **25**, 263.

Parker, J. G. (1959). *Phys. Fluids*, **2**, 449.

Parker, J. G. (1961). *J. Chem. Phys.*, **34**, 1763.

Parker, J. G. (1964). *J. Chem. Phys.*, **41**, 1600.

Parker, R. C., K. R. Applegate and L. J. Slutsky (1966). *J. Phys. Chem.*, **70**, 3018.

Parker, R. C., L. J. Slutsky and K. R. Applegate (1968). *J. Phys. Chem.*, **72**, 3177.

Parpiev, K., P. K. Khabibullaev, M. G. Khaliulin and M. I. Shakhparonov (1970). *Sov. Phys. Acoust.*, **15**, 345. [(1969). *Akust. Zh.*, **15**, 401.]

Parpiev, K., P. K. Khabibullaev and M. G. Khaliulin (1970). *Sov. Phys. Acoust.*, **15**, 406. [(1969). *Akust. Zh.*, **15**, 466.]

Pauling, L. and E. B. Wilson (1935). *Introduction to Quantum Mechanics*, McGraw–Hill, New York.

Pedinoff, M. E. (1959). University of California, Dept. of Physics, Tech. Report No. 14. [Quoted by Subrahmanyam and Piercy (1965b).]

Pedinoff, M. E. (1962). *J. Chem. Phys.*, **36**, 777.

Pellam, J. R. and J. K. Galt (1946). *J. Chem. Phys.*, **14**, 608.

Pentin, Y. A., Z. Sharipov, G. G. Kotova, A. V. Kamernitskii and A. A. Akhrem (1963). *Zh. Strukt. Khim.*, **4**, 194.

Peterlin, A. (1967). *J. Polymer Sci. A2*, **5**, 179.

Peterlin, A. and C. Reinhold (1967). *Trans. Soc. Rheology*, **11**, 15.

Pethrick, R. A. and E. Wyn-Jones (1968). *J. Chem. Phys.*, **49**, 5349.

Pethrick, R. A. and E. Wyn-Jones (1969). *J. Chem. Soc. A*, 713.

Petrauskas, A. A. (1959). Unpublished work. [Quoted by Lamb (1965).]

Philippoff, W. (1934). *Phys. Z.*, **35**, 900.

Philippoff, W. (1954). *J. Appl. Phys.*, **25**, 1102.

Philippoff, W. (1963). *J. Appl. Phys.*, **34**, 1507.

Philippoff, W. (1964). *Trans. Soc. Rheology*, **8**, 117.

Phillips, M. C. (1969). Thesis, University of Glasgow.

Piccirelli, R. and T. A. Litovitz (1957). *J. Acoust. Soc. Am.*, **29**, 1009.

Pierce, G. W. (1925). *Proc. Am. Acad. Arts Sci.*, **60**, 271.

Piercy, J. E. and J. Lamb (1956). *Trans. Faraday Soc.*, **52**, 930.

Piercy, J. E. (1961). *J. Acoust. Soc. Am.*, **33**, 198.

Piercy, J. E. and S. V. Subrahmanyam (1965a). *J. Chem. Phys.*, **42**, 1475.

Piercy, J. E. and S. V. Subrahmanyam (1965b). *J. Chem. Phys.*, **42**, 4011.

Piercy, J. E. (1965). *J. Chem. Phys.*, **43**, 4066.

Piercy, J. E. and M. G. S. Rao (1967a). *J. Chem. Phys.*, **46**, 3951.

Piercy, J. E. and M. G. S. Rao (1967b). *J. Acoust. Soc. Am.*, **41**, 1063.

Piercy, J. E. and M. G. S. Rao (1967c). *J. Acoust. Soc. Am.*, **41**, 1591.

Pigott, M. T. and R. C. Strum (1967a). *J. Acoust. Soc. Am.*, **41**, 662.

Pigott, M. T. and R. C. Strum (1967b). *Rev. Sci. Inst.*, **38**, 743.

Pinkerton, J. M. M. (1947). *Nature*, **160**, 128.

Pinkerton, J. M. M. (1949). *Proc. Phys. Soc. B*, **62**, 129, 286.

Pinnow, D. A., S. J. Candau, J. T. LaMacchia and T. A. Litovitz (1968). *J. Acoust. Soc. Am.*, **43**, 131.

Pippard, A. B. (1955). *Phil. Mag.*, **46**, 1104.

Plass, K. G. (1965). *Acustica*, **15**, 446.

Plass, K. G. and A. Kehl (1968). *Acustica*, **20**, 360.

Plazek, D. J. and J. D. Ferry (1956). *J. Phys. Chem.*, **60**, 289.

Popov, E. D. and V. F. Yakovlev (1969). *Sov. Phys. Acoust.*, **15**, 118. [*Akust. Zh.*, **15**, 138.]

Pryde, J. A. (1966). *The Liquid State*, Hutchinson, London.

Ptitsyn, O. B. and Y. E. Eisner (1958). *Zh. Fiz. Khim.*, **32**, 2464.

Ptitsyn, O. B. and Y. E. Eisner (1959). *Sov. Phys. Tech. Phys.*, **4**, 1020. [*Zh. Tekh. Fiz.*, **29**, 1117.]

Purdie, N. and C. A. Vincent (1967). *Trans. Faraday Soc.*, **63**, 2745.

Raff, L. M. and T. G. Winter (1968). *J. Chem. Phys.*, **48**, 3992.

Rao, M. R. (1940). *Indian J. Phys.*, **14**, 109.

Rao, M. R. (1941). *J. Chem. Phys.*, **9**, 682.

Rapp, D. and T. E. Sharp (1963). *J. Chem. Phys.*, **38**, 2641.

Rapuano, R. A. (1950). Massachusetts Institute of Technology, Research Lab. of Electronics, Report No. 151. [Quoted by Lamb and Andreae (1951).]

Rasmussen, R. A. (1967). *J. Chem. Phys.*, **46**, 211.

Rasmussen, R. A. (1968). *J. Chem. Phys.*, **48**, 3364.

Rayleigh, Lord (1877). *The Theory of Sound*, Macmillan, London.

Rayleigh, Lord (1899). *Phil. Mag.*, **47**, 308.

Read, A. W. (1968). *Adv. Mol. Relax. Proc.*, **1**, 257.

Ree, T. S., T. Ree and H. Eyring (1962). *Proc. Nat. Acad. Sci. U.S.*, **48**, 501.

Reeves, L. W. and K. O. Strømme (1960). *Canad. J. Chem.*, **38**, 1241.

Richards, J. R., K. Ninomiya and J. D. Ferry (1963). *J. Phys. Chem.*, **67**, 323.

Richardson, E. G. (1962). *Ultrasonic Physics*, Elsevier, Amsterdam.

Roberts, C. S. (1963). *Phys. Rev.*, **131**, 209.

Roesler, H. and K. F. Sahm (1965). *J. Acoust. Soc. Am.*, **37**, 386.

Rossing, T. D. and S. Legvold (1955). *J. Chem. Phys.*, **23**, 1118.

Rouse, P. E. (1953). *J. Chem. Phys.*, **21**, 1272.

Rouse, P. E. and K. Sittel (1953). *J. Appl. Phys.*, **24**, 690.

Rowlinson, J. S. (1969). *Liquids and Liquid Mixtures*, Butterworths, London.

Saito, N., K. Okano, S. Iwayanagi and T. Hideshima (1963). *Solid State Phys.*, **14**, 343.

Sakanishi, A. (1968). *J. Chem. Phys.*, **48**, 3850.

Sanders, J. F., J. D. Ferry and R. H. Valentine (1968). *J. Polymer Sci. A2*, **6**, 967.

Sather, N. F. and J. S. Dahler (1961). *J. Chem. Phys.*, **35**, 2029.

Sather, N. F. and J. S. Dahler (1962). *J. Chem. Phys.*, **37**, 1947.

Sato, Y., S. T. Tsuchiya and K. Kuratani (1969). *J. Chem. Phys.*, **50**, 1911.

Saunders, G. A., N. G. Pace and T. Alper (1967). *Nature*, **216**, 1298.

Saunders, P. R., D. M. Stern, S. F. Kurath, C. Sakoonkim and J. D. Ferry (1959). *J. Colloid Sci.*, **14**, 222.

Sawyer, W. M. and J. D. Ferry (1950). *J. Am. Chem. Soc.*, **72**, 5030.

Schaaffs, W. and C. Kalweit (1960). *Acustica*, **10**, 385.

Schaaffs, W. (1963). *Molekularakustik*, Springer, Berlin.

Scheraga, H. A. (1955). *J. Chem. Phys.*, **23**, 1526.

Schnaus, U. E. (1965). *J. Acoust. Soc. Am.*, **37**, 1.

Schneider, W. G. (1951). *Canad. J. Chem.*, **29**, 243.

Schwartz, R. N., Z. I. Slawsky and K. F. Herzfeld (1952). *J. Chem. Phys.*, **20**, 1591.

Schwartz, R. N. and K. F. Herzfeld (1954). *J. Chem. Phys.*, **22**, 767.

Schwarzl, F. R. and L. C. E. Struik (1968). *Adv. Mol. Relax. Proc.*, **1**, 201.

Scott, G. D. (1962). *Nature*, **194**, 956.

Seeger, A., H. Donth and F. Pfaff (1957). *Disc. Faraday Soc.*, **23**, 19.

Sell H. (1937). *Z. Tech. Phys.*, **18**, 3.

Sette, D., A. Busala and J. C. Hubbard (1955). *J. Chem. Phys.*, **23**, 787.

Sharma, R. D. (1969). *J. Chem. Phys.*, **50**, 919.

Sharma, R. D. and C. A. Brau (1969). *J. Chem. Phys.*, **50**, 924.

Sheppard, N. and G. J. Szasz (1949). *J. Chem. Phys.*, **17**, 86.

Sheppard, N. (1959). *Adv. Spectr.*, **1**, 288.

Shields, F. D. and R. T. Lagemann (1957). *J. Acoust. Soc. Am.*, **29**, 470.

Shields, F. D. (1960). *J. Acoust. Soc. Am.*, **32**, 180.

Shields, F. D. (1962). *J. Acoust. Soc. Am.*, **34**, 271.

Shields, F. D., K. P. Lee and W. J. Wiley (1965). *J. Acoust. Soc. Am.*, **37**, 724.

Shields, F. D. and J. A. Burks (1968). *J. Acoust. Soc. Am.*, **43**, 510.

Shields, F. D. (1969). *J. Acoust. Soc. Am.*, **45**, 481.

Shields, F. D. and J. Faughn (1969). *J. Acoust. Soc. Am.*, **46**, 158.

Shin, H. K. (1967a). *J. Chem. Phys.*, **46**, 3688.

Shin, H. K. (1967b). *J. Chem. Phys.*, **47**, 3302.

Shin, H. K. (1968). *J. Am. Chem. Soc.*, **90**, 3025, 3029.

Simmons, J. H. and P. B. Macedo (1968). *J. Acoust. Soc. Am.*, **43**, 1295.

Simmons, J. M. (1966). *J. Sci. Inst.*, **43**, 887.

Simpson, C. J. S. M., T. R. D. Chandler and A. C. Strawson (1969). *J. Chem. Phys.*, **51**, 2214.

Singh, R. P., G. S. Darbari and G. S. Verma (1967). *J. Chem. Phys.*, **46**, 151.

Singh, R. P. and G. S. Verma (1968). *J. Phys. C.*, **1**, 1476.

Sinnott, K. M. (1960). *J. Polymer Sci.*, **42**, 3.

Sittel, K., P. E. Rouse and E. D. Bailey (1954). *J. Appl. Phys.*, **25**, 1312.

Sittig, E. (1960). *Acustica*, **10**, 81.

Slie, W. M. and T. A. Litovitz (1961). *J. Acoust. Soc. Am.*, **33**, 1412.

Slie, W. M. and T. A. Litovitz (1963). *J. Chem. Phys.*, **39**, 1538.

Slie, W. M., A. R. Donfor and T. A. Litovitz (1966). *J. Chem. Phys.*, **44**, 3712.

Smith, D. H. and R. G. Harlow (1963). *Brit. J. Appl. Phys.*, **14**, 102.

Solovyev, V. A., C. J. Montrose, M. H. Watkins and T. A. Litovitz (1968). *J. Chem. Phys.*, **48**, 2155.

Splitt, G. (1969). *Z. Phys.*, **225**, 60.

Starunov, V. S., E. V. Tiganov and I. L. Fabelinskii (1966). *J.E.T.P. Lett.*, **4**, 176. [*Zh. Eksp. Teor. Fiz. Pisma*, **4**, 262.]

Staveley, L. A. K., K. R. Hart and W. I. Tupman (1953). *Disc. Faraday Soc.*, **15**, 130.

Stegeman, G. I. A. and B. P. Stoicheff (1968). *Phys. Rev. Lett.*, **21**, 202.

Stephens, R. W. B. and A. E. Bate (1966). *Acoustics and Vibrational Physics*, Arnold, London.

Stevens, B. (1967). *Collisional Activation in Gases*, Pergamon, Oxford.

Stewart, E. S. (1946). *Phys. Rev.*, **69**, 632.

Stewart, E. S. and J. L. Stewart (1963). *J. Acoust. Soc. Am.*, **35**, 975.

Stewart, J. L. (1946). *Rev. Sci. Inst.*, **17**, 59.

Stewart, J. L. and E. S. Stewart (1952). *J. Acoust. Soc. Am.*, **24**, 22.

Stokes, G. S. (1845). *Trans. Camb. Phil. Soc.*, **8**, 287.

Strakna, R. E. and H. T. Savage (1964). *J. Appl. Phys.*, **35**, 1445.

Strauch, J. G. and J. C. Decius (1966). *J. Chem. Phys.*, **44**, 3319.

Stretton, J. L. (1965). *Trans. Faraday Soc.*, **61**, 1053.

Struykov, V. B. and I. F. Shchegolev (1967). *Cryogenics*, **7**, 302.

Stuehr, J., E. Yeager, T. Sachs and F. Hovorka (1963). *J. Chem. Phys.*, **38**, 587.

Stuehr, J. and E. Yeager (1967). *J. Chem. Phys.*, **46**, 1222.

Subrahmanyam, S. V. and J. E. Piercy (1965a). *J. Acoust. Soc. Am.*, **37**, 340.

Subrahmanyam, S. V. and J. E. Piercy (1965b). *J. Chem. Phys.*, **42**, 1845.

Swift, T. J. and R. E. Connick (1962). *J. Chem. Phys.*, **37**, 307.

Swift, T. J. and R. E. Connick (1964). *J. Chem. Phys.*, **41**, 2553.

Szasz, G. J., N. Sheppard and D. H. Rank (1948). *J. Chem. Phys.*, **16**, 704.

Tabuchi, D. (1958). *J. Chem. Phys.*, **28**, 1014.

Tabuchi, D. (1968). *Rep. 6th Int. Cong. Acoust.*, J-25.

Takayanagi, K. (1952). *Prog. Theo. Phys. Japan*, **8**, 111.

Takayanagi, K. (1957). *Proc. Phys. Soc. A*, **70**, 348.

Takayanagi, K. (1965). *Adv. At. Mol. Phys.*, **1**, 149.

Tamm, K., G. Kurtze and R. Kaiser (1954). *Acustica*, **4**, 380.

Tamm, K. (1968). *Rep. 6th Int. Cong. Acoust.*, GP-25.

Tammann, G. and W. Hesse (1926). *Z. Anorg. Allgem. Chem.*, **156**, 245.

Tanaka, H., A. Sakanishi, M. Kaneko and J. Furuichi (1966). *J. Polymer Sci. C*, **15**, 317.

Tanaka, H. and A. Sakanishi (1969). *Proc. 5th Int. Cong. Rheology*, in the press.

Tanczos, F. I. (1956). *J. Chem. Phys.*, **25**, 439.

Tatsumoto, N. (1967). *J. Chem. Phys.*, **47**, 4561.

Tauke, J., T. A. Litovitz and P. B. Macedo (1968). *J. Am. Ceram. Soc.*, **51**, 158.

Taylor, R. G. F. and A. J. Pointon (1969). *Cont. Phys.*, **10**, 159.

Taylor, R. L. and S. Bitterman (1969). *J. Chem. Phys.*, **50**, 1720.

Teutonico, L. J. (1958). Thesis, Brown University. [Quoted by Truell, Elbaum and Chick (1969).]

Thomas, T. H., E. Wyn-Jones and W. J. Orville-Thomas (1969). *Trans. Faraday Soc.*, **65**, 974.

Thurston, G. B. and A. Peterlin (1967). *J. Chem. Phys.*, **46**, 4881.

Thurston, G. B. and J. L. Schrag (1966). *J. Chem. Phys.*, **45**, 3373.

Tielsch, H. and H. Tanneberger (1954). *Z. Phys.*, **137**, 256.

Tittman, B. R. and H. E. Bömmel (1967). *Rev. Sci. Inst.*, **38**, 1491.

Tittman, B. R. and H. E. Bömmel (1968). *Rev. Sci. Inst.*, **39**, 614.

Tobolsky, A. V. and J. R. McLoughlin (1952). *J. Polymer Sci.*, **8**, 543.

Truell, R., C. Elbaum and B. B. Chick (1969). *Ultrasonic Methods in Solid State Physics*, Academic Press, New York.

Truesdell, C. (1953). *J. Rat. Mech. Anal.*, **2**, 594.

Tschoegl, N. W. (1963). *J. Chem. Phys.*, **39**, 149.

Tschoegl, N. W. and J. D. Ferry (1963). *Kolloid Z.*, **189**, 37.

Tschoegl, N. W. (1964). *J. Chem. Phys.*, **40**, 473.

Tschoegl, N. W. and J. D. Ferry (1964a). *J. Phys. Chem.*, **68**, 867.

Tschoegl, N. W. and J. D. Ferry (1964b). *J. Am. Chem. Soc.*, **86**, 1474.

Tyndall, J. (1881). *Proc. Roy. Soc. A*, **31**, 307.

Unland, M. L. and W. H. Flygare (1966). *J. Chem. Phys.*, **45**, 2421.

Valley, L. M. and R. C. Amme (1969). *J. Chem. Phys.*, **50**, 3190.

Venketeswaran, C. S. (1942). *Proc. Indian Acad. Sci. A*, **15**, 322, 371.

Verlet, L. (1967). *Phys. Rev.*, **159**, 98.

Victor, A. E. and R. T. Beyer (1970). *J. Chem. Phys.*, **52**, 1573.

Vigoureux, P. (1950). *Ultrasonics*, Chapman and Hall, London.

Vogel, H. (1921). *Phys. Z.*, **22**, 645.

Volterra, V. (1969). *Phys. Rev.*, **180**, 156.

Voronel, A. V., Y. R. Chashkin, V. A. Popov and V. G. Simkin (1964). *Sov. Phys. J.E.T.P.*, **18**, 568. [(1963). *Zh. Eksp. Teor. Fiz.*, **45**, 828.]

Wang Chang, C. S. and G. E. Uhlenbeck (1951). Univ. Michigan, Report CM-681. [Quoted by Wang Chang, Uhlenbeck and de Boer (1964).]

Wang Chang, C. S., G. E. Uhlenbeck and J. de Boer (1964). In J. de Boer and G. E. Uhlenbeck (Eds.), *Studies in Statistical Mechanics*, Vol. 2, North Holland, Amsterdam.

Wauk, M. T. and D. K. Winslow (1968). *Appl. Phys. Lett.*, **13**, 286.

Weinreich, G., T. M. Sanders and H. G. White (1959). *Phys. Rev.*, **114**, 33.

White, D. R. and R. C. Millikan (1963). *J. Chem. Phys.*, **39**, 1803.

Widom, B. (1960). *J. Chem. Phys.*, **32**, 913.

Wight, H. M. (1956). *J. Acoust. Soc. Am.*, **28**, 459.

Willbourn, A. H. (1958). *Trans. Faraday Soc.*, **54**, 717.

Williams, E. J., T. H. Thomas, E. Wyn-Jones and W. J. Orville-Thomas (1968). *J. Mol. Struct.*, **2**, 307.

Williams, G. and D. C. Watts (1970). *Trans. Faraday Soc.*, **66**, 80.

Williamson, R. C. (1969). *Rev. Sci. Inst.*, **40**, 666.

Winter, T. G. (1963). *J. Acoust. Soc. Am.*, **35**, 1882.

Winter, T. G. and G. L. Hill (1967). *J. Acoust. Soc. Am.*, **42**, 848.

Wolf, P. (1965). *Acustica*, **15**, 39.

Wray, K. L. (1962). *J. Chem. Phys.*, **36**, 2597.

Wyn-Jones, E. (1963). Thesis, University of Wales.

Wyn-Jones, E. and W. J. Orville-Thomas (1966). *Chem. Soc. Spec. Publ.*, **20**, 209.

Wyn-Jones, E. and W. J. Orville-Thomas (1968). *Trans. Faraday Soc.*, **64**, 2907.

Yager, W. (1936). *J. Appl. Phys.*, **7**, 434.

Yamada, K. and Y. Fujii (1966). *J. Acoust. Soc. Am.*, **39**, 250.

Yardley, J. T. and C. B. Moore (1968). *J. Chem. Phys.*, **49**, 1111.

Yardley, J. T., M. N. Fertig and C. B. Moore (1970). *J. Chem. Phys.*, **52**, 1450.

Yasunaga, T., N. Tatsumoto and M. Miura (1964). *Bull. Chem. Soc. Japan*, **37**, 1655.

Yasunaga, T., N. Tatsumoto and M. Miura (1965). *J. Chem. Phys.*, **43**, 2735.

Yasunaga, T., M. Tanoura and M. Miura (1965). *J. Chem. Phys.*, **43**, 3512.

Yasunaga, T., H. Oguri and M. Miura (1967). *J. Colloid Sci.*, **23**, 352.

Yasunaga, T., S. Fujii and M. Miura (1968). *Rep. 6th Int. Cong. Acoust.*, J-61.

Yasunaga, T., N. Tatsumoto, H. Inoue and M. Miura (1969). *J. Phys. Chem.*, **73**, 477.

Yeager, E., J. Bugosh, F. Hovorka and J. McCarthy (1949). *J. Chem. Phys.*, **17**, 411.

Yeager, E., H. Dietrick and F. Hovorka (1953). *J. Acoust. Soc. Am.*, **25**, 456.

Young, J. M. and A. A. Petrauskas (1956). *J. Chem. Phys.*, **25**, 943.

Yun, S. S. and R. T. Beyer (1964). *J. Chem. Phys.*, **40**, 2538.

Zana, R. and E. Yeager (1967a). *J. Phys. Chem.*, **71**, 521.

Zana, R. and E. Yeager (1967b). *J. Phys. Chem.*, **71**, 3502.

Zana, R. and J. Lang (1968a). *Compt. Rend. C*, **266**, 893.

Zana, R. and J. Lang (1968b). *Compt. Rend. C*, **266**, 1347.

Zeleznik, F. J. (1967). *J. Chem. Phys.*, **47**, 3410.

Zener, C. (1931). *Phys. Rev.*, **37**, 556.

Zener, C. (1938). *Phys. Rev.*, **53**, 90.

Zener, C. (1940). *Proc. Phys. Soc.*, **52**, 152.

Zimm, B. H. (1956). *J. Chem. Phys.*, **24**, 269.

Zwanzig, R. (1961). *J. Chem. Phys.*, **34**, 1931.

Zwanzig, R. and R. D. Mountain (1965). *J. Chem. Phys.*, **43**, 4464.

AUTHOR INDEX

Page numbers in *italics* indicate that an author's name is not directly quoted.

SUBJECT INDEX